Developmental Programming of Cardiovascular Disease

Colloquium
Digital Library of Life Sciences

This e-book is a copyrighted work in the Colloquium Digital Library—an innovative collection of time saving references and tools for researchers and students who want to quickly get up to speed in a new area or fundamental biomedical/life sciences topic. Each PDF e-book in the collection is an in-depth overview of a fast-moving or fundamental area of research, authored by a prominent contributor to the field. We call these e-books *Lectures* because they are intended for a broad, diverse audience of life scientists, in the spirit of a plenary lecture delivered by a keynote speaker or visiting professor. Individual e-books are published as contributions to a particular thematic **series**, each covering a different subject area and managed by its own prestigious editor, who oversees topic and author selection as well as scientific review. Readers are invited to see highlights of fields other than their own, keep up with advances in various disciplines, and refresh their understanding of core concepts in cell & molecular biology.

For the full list of published and forthcoming Lectures, please visit the Colloquium homepage: www.morganclaypool.com/page/lifesci

Access to Colloquium Digital Library is available by institutional license. Please e-mail info@morganclaypool.com for more information.

Morgan & Claypool Life Sciences is a signatory to the STM Permission Guidelines. All figures used with permission.

Colloquium Series on Integrated Systems Physiology: From Molecule to Function to Disease

Editors

D. Neil Granger, *Louisiana State University Health Sciences Center*

Joey P. Granger, *University of Mississippi Medical Center*

Physiology is a scientific discipline devoted to understanding the functions of the body. It addresses function at multiple levels, including molecular, cellular, organ, and system. An appreciation of the processes that occur at each level is necessary to understand function in health and the dysfunction associated with disease. Homeostasis and integration are fundamental principles of physiology that account for the relative constancy of organ processes and bodily function even in the face of substantial environmental changes. This constancy results from integrative, cooperative interactions of chemical and electrical signaling processes within and between cells, organs and systems. This eBook series on the broad field of physiology covers the major organ systems from an integrative perspective that addresses the molecular and cellular processes that contribute to homeostasis. Material on pathophysiology is also included throughout the eBooks. The state-of the-art treatises were produced by leading experts in the field of physiology. Each eBook includes stand-alone information and is intended to be of value to students, scientists, and clinicians in the biomedical sciences. Since physiological concepts are an ever-changing work-in-progress, each contributor will have the opportunity to make periodic updates of the covered material.

Published titles

(for future titles please see the website, www.morganclaypool.com/page/lifesci)

Developmental Programming of Cardiovascular Disease
Barbara T. Alexander
www.morganclaypool.com

ISBN: 9781615046027   paperback

ISBN: 9781615046034   ebook

DOI: 10.4199/C00084ED1V01Y201305ISP038

A Publication in the

COLLOQUIUM SERIES ON INTEGRATED SYSTEMS PHYSIOLOGY: FROM MOLECULE TO FUNCTION TO DISEASE

Lecture #38

Series Editors: D. Neil Granger, LSU Health Sciences Center, and Joey P. Granger, University of Mississippi Medical Center

Series ISSN

ISSN 2154-560X   print
ISSN 2154-5626   electronic

Developmental Programming of Cardiovascular Disease

Barbara T. Alexander
Department of Physiology and Biophysics
University of Mississippi Medical Center

COLLOQUIUM SERIES ON INTEGRATED SYSTEMS PHYSIOLOGY:
FROM MOLECULE TO FUNCTION TO DISEASE #38

 MORGAN&CLAYPOOL LIFE SCIENCES

ABSTRACT

Numerous epidemiological studies report that birth weight is inversely associated with blood pressure, suggesting that slow growth during fetal life programs hypertension and increased risk for cardiovascular disease in later life. Different experimental models are used to provide proof of concept for the theory of developmental programming of cardiovascular disease, and studies in these different animal models are providing insight into the etiology of chronic disease programmed by an imbalance in nutrition during early life or exposure to maternal complications during pregnancy. Alterations in the regulatory systems key to the long-term control of blood pressure are implicated in the etiology of hypertension that results from adverse exposures during early development. Epigenetic processes are also implicated in the increased risk for programmed cardiovascular disease and the passage of programmed cardiovascular risk to the next generation. Sex, age, and early postnatal growth impact later programmed risk; programmed risk is also amplified in response to a secondary challenge that includes normal physiological processes such as pregnancy. Thus, this book will highlight how events during early life impact later cardiovascular health in a manner that is sex- and age-dependent and can be transmitted to the next generation.

KEYWORDS

developmental programming, fetal programming, low birth weight, intrauterine growth restriction, preterm birth, hypertension, nephron number, sex differences, cardiovascular disease, aging

Contents

CHAPTER 1

Introduction

The concept of the Developmental Origins of Health and Disease hypothesizes that influences during critical periods of fetal or early postnatal life program permanent adaptive structural, physiological, and epigenetic modifications resulting in the development of cardiovascular disease in later life. Low birth weight, a surrogate marker of poor fetal growth, is linked to increased risk for heart disease, hypertension, and stroke. Numerous animal models support this association and provide insight into the mechanisms by which adverse influences during early life can program increased cardiovascular risk in later life. The importance of the kidney, altered vascular function, the renin angiotensin system, the sympathetic nervous system, and oxidative stress are implicated in the etiology of programmed cardiovascular risk. Sex differences in response to fetal insult are observed in many animal models of developmental programming with phenotypic outcome linked to the severity of the insult. Population and animal studies indicate that the association between birth weight and blood pressure amplifies with age. Thus, sex and age may impact the progression and severity of chronic disease induced by exposure to an adverse insult during early life. Transmission of heritable risk to the next generation suggests that epigenetic modifications may contribute to the passage of programmed risk to subsequent generations. Future research is needed to understand the underlying mechanisms of developmental programming of chronic disease in order to address therapeutic options for men and women at risk. This book will highlight the historical perspective, the strong link between birth weight, and later cardiovascular disease provided by epidemiological studies, the different experimental models utilized for study of potential mechanisms, and the potential mechanistic pathways that contribute to the developmental programming of cardiovascular risk.

· · · ·

CHAPTER 2

Historical Perspective

Poor socioeconomic status is linked to an increase in coronary heart disease [1, 2]. Known cardio-vascular risk factors may contribute [3], yet the causative factors are not clear. In 1973, Forsdahl noted that poor socioeconomic conditions during childhood and adolescence were associated with increased mortality in adulthood from chronic disease [4], suggesting that adverse influences during early life could impact later chronic health. In 1985, Wadsworth and others reported that blood pressure measured in a national cohort at 36 years of age was related to socioeconomic factors with birth weight as a contributory factor [5]. In 1986, Barker and Osmond reported that the geographical distribution of death from coronary heart disease closely resembled the geographical pattern of infant mortality for that same generation [6] (Figure 1). Barker and Osmond also examined the

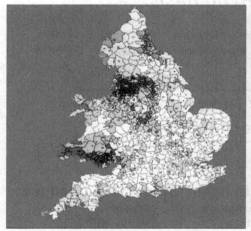

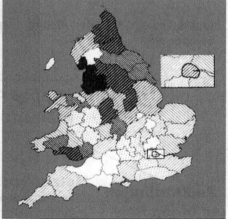

Standardized mortality ratios (SMR) for coronary heart disease in England and Wales among men aged 35-74 years during 1968-1978. (Fig 1.1)

Infant mortality rates per 1000 births in England and Wales during 1901-10. (Fig 1.2)

FIGURE 1: Standardized mortality ratios (SMR) for coronary heart disease in England and Wales among men aged 35–74 years during 1968–1978 (Figure 1.1) and infant mortality rates per 1000 births in England and Wales during 1901–2010 (Figure 1.2). Used with permission from the Second Edition of *Mother's, Babies and Health in Later Life published in 1998* (Figures 1.2 and 1.3).

link between birth weight and blood pressure and reported that birth weight was inversely related to blood pressure in a cohort of children studied at 10 years of age [7]. This observation was important as it demonstrated an association between birth weight and blood pressure prior to the development of adult lifestyle risk factors that could confound this relationship. Based on these findings, they concluded that influences during fetal life that lead to increased infant mortality might also contribute to increased cardiovascular risk in later life. They expanded their investigation in additional studies to examine the association between birth weight and blood pressure in adults [8] and concluded that hypertension may be the link between an adverse fetal environment and later cardiovascular risk.

2.1   BARKER HYPOTHESIS AND FETAL PROGRAMMING OF CARDIOVASCULAR DISEASE

Low birth weight, or 5.5 pounds or less, serves a crude marker for poor fetal growth. Fetal growth is dependent upon adequate delivery of nutrients and oxygen to the fetus [9]. Based on the inverse relationship between birth weight and blood pressure and findings related to the geographical distri-

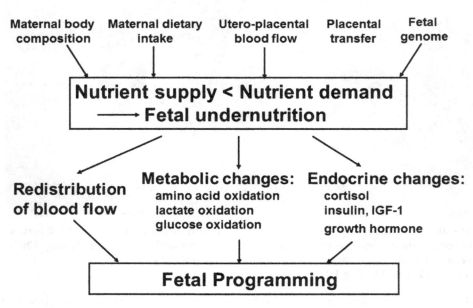

FIGURE 2: Fetal adaptations to undernutrition: a framework. Used with permission from the Second Edition of *Mother's, Babies and Health in Later Life published in 1998* (Figure 8.2).

bution of coronary heart disease and infant mortality within the same generation, Barker proposed the theory of fetal programming of adult disease [10]. He hypothesized that when nutrient supply is less than nutrient demand during fetal life, the fetus will respond to a reduction in nutrients by sparing the brain at the expense of other organs during development in order to survive to birth [10] (Figure 2). He further hypothesized that the reduction in blood supply containing nutrients and oxygen needed for proper development during fetal life leads to impaired organogenesis if it occurs during a key period of development. The theory of fetal programming, also referred to as the Barker Hypothesis, proposes that impaired organogenesis in response to fetal undernutrition programs changes in structure, endocrine, and physiological function of the fetus, leading to an increase in blood pressure and cardiovascular risk in later life [11].

2.2   POSTNATAL INFLUENCES AND THE DEVELOPMENTAL PROGRAMMING OF CARDIOVASCULAR DISEASE

The Barker hypothesis proposed that undernutrition during fetal life programs an increase in risk factors for cardiovascular disease 9 [10]. Yet, accelerated weight gain during early postnatal life also impacts risk factors for cardiovascular disease [12, 13], suggesting that the period of developmental vulnerability includes not only fetal life but also early postnatal life. Experimental studies support the importance of accelerated growth during early life on later cardiovascular risk [14], leading to expansion of the theory from fetal programming to the Developmental Origins of Health and Disease (DOHaD). In addition, the impact of adverse influences during early postnatal life may have a long-term impact on later chronic health in the absence of impaired growth during fetal life [15].

· · · · ·

CHAPTER 3

Epidemiological Studies

3.1 COMPLICATIONS DURING PREGNANCY, BIRTH WEIGHT, AND LATER CARDIOVASCULAR RISK

Pregnancy is associated with numerous physiological adaptations that occur to meet the demands of the developing fetus [16]. Complications during pregnancy that impact nutrient delivery to the fetus can slow fetal growth resulting in intrauterine growth restriction (IUGR). Maternal complications that impact fetal growth include hypertension, diabetes, smoking, and obesity [17]. Maternal age and nutrition in addition to poor prenatal care can also contribute to fetal growth restriction [17]. Numerous studies indicate that complications during pregnancy not only compromise the growth and health of the fetus but also program an increase in cardiovascular risk in offspring that is present in childhood and persist into adulthood.

Numerous studies link complications during pregnancy with later cardiovascular risk in the offspring. Blood pressure is increased in children and young adults of preeclamptic pregnancies [18, 19, 20] (Figure 3). Cardiovascular risk is also increased in children exposed to smoking [19],

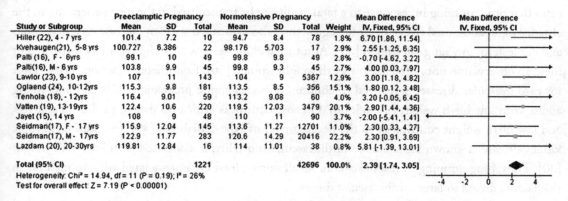

Study or Subgroup	Preeclamptic Pregnancy			Normotensive Pregnancy			Weight	Mean Difference IV, Fixed, 95% CI	Mean Difference IV, Fixed, 95% CI
	Mean	SD	Total	Mean	SD	Total			
Hiller (22), 4 - 7 yrs	101.4	7.2	10	94.7	8.4	78	1.8%	6.70 [1.86, 11.54]	
Kvehaugen(21), 5-8 yrs	100.727	6.386	22	98.176	5.703	17	2.9%	2.55 [-1.25, 6.35]	
Palti (16), F - 6yrs	99.1	10	49	99.8	9.8	49	2.8%	-0.70 [-4.62, 3.22]	
Palti(16), M - 6 yrs	103.8	9.9	45	99.8	9.3	45	2.7%	4.00 [0.03, 7.97]	
Lawlor (23), 9-10 yrs	107	11	143	104	9	5367	12.9%	3.00 [1.18, 4.82]	
Oglaend (24), 10-12yrs	115.3	9.8	181	113.5	8.5	356	15.1%	1.80 [0.12, 3.48]	
Tenhola (18), - 12yrs	116.4	9.01	59	113.2	9.08	60	4.0%	3.20 [-0.05, 6.45]	
Vatten (19), 13-19yrs	122.4	10.6	220	119.5	12.03	3479	20.1%	2.90 [1.44, 4.36]	
Jayet (15), 14 yrs	108	9	48	110	11	90	3.7%	-2.00 [-5.41, 1.41]	
Seidman(17), F - 17 yrs	115.9	12.04	145	113.6	11.27	12701	11.0%	2.30 [0.33, 4.27]	
Seidman(17), M - 17yrs	122.9	11.77	283	120.6	14.29	20416	22.2%	2.30 [0.91, 3.69]	
Lazdam (20), 20-30yrs	119.81	12.84	16	114	11.01	38	0.8%	5.81 [-1.39, 13.01]	
Total (95% CI)			1221			42696	100.0%	2.39 [1.74, 3.05]	
Heterogeneity: Chi² = 14.94, df = 11 (P = 0.19); I² = 26%									
Test for overall effect: Z = 7.19 (P < 0.00001)									

FIGURE 3: Mean difference in systolic pressure in mmHg between those who were exposed to preeclampsia in utero and controls. Used with permission from *Pediatrics*. 2012; 129: e1552–1561 (Figure 2).

maternal diabetes [21], and obesity [22]. Hypertension is a major risk factor for later cardiovascular disease and blood pressure during childhood is a strong predictor of blood pressure in later life [23]. Low birth weight, weight at birth of 5.5 pounds or less, serves as a crude marker for slow fetal growth. Numerous cross-sectional studies indicate that birth weight is inversely related to blood pressure during childhood [24, 25]. Longitudinal studies support this observation [26] and demonstrate that the inverse relationship between birth weight and blood pressure extends well into adulthood [27]. Age impacts this relationship and by 60 years of age systolic blood pressure is decreased by 5.2 mm Hg for every kg decrease in birth weight [28]. Growth during infant life can impact later cardiovascular health [29] and the association between birth weight and blood pressure persists even when adjusted for confounding factors such as body mass index (BMI) [25]. Variability in blood pressure is another risk factor for later increased cardiovascular risk [30]. Birth weight is not only inversely correlated to blood pressure but also impacts the magnitude of fluctuations (within-individual variability) in blood pressure [31]. Thus, long-term variability in blood pressure is increased in children with low birth weight [31] and represents a major risk factor for later cardiovascular disease.

In addition to the inverse association with blood pressure, birth weight is also linked to numerous other risk factors for cardiovascular disease. Excessive weight gain or obesity is a major risk factor for hypertension and renal disease [32]. Children and young adults of pregnancies complicated by preeclampsia exhibit a marked increase in body mass index (BMI) [18]. Increased weight and fat deposition is observed in low birth weight women born to pregnancies impacted by the severe famine during the Dutch Hunger winter of 1944–1945 [33]. In addition, despite no complications during their pregnancy, birth weight is reduced in offspring of women born with low birth weight due to gestational exposure to the Dutch Hunger winter [34]. Thus, this study indicates that programming in response to a fetal insult can be transmitted to the next generation in the absence of an additional stress. Slow fetal growth is also associated with an increase in triglycerides and cholesterol in young adulthood [35]. Adult total cholesterol is also higher in low birth weight men, an observation not noted for low birth weight women [36]. Endothelial function, a risk factor for cardiovascular disease, is reduced in children of preeclamptic pregnancies [20, 37] and young adults with low birth weight [20, 35, 38]. Glucose control is critical to overall cardiovascular risk and low birth weight contributes to the development of type-2 diabetes [39]. In addition, chronic kidney disease, a known associate for cardiovascular mortality, is associated with low birth weight [40]. Thus, programming in response to an insult during fetal life is associated with an increase in risk factors linked to later cardiovascular disease.

Preterm birth in addition to slow fetal growth can lead to low birth weight. Preterm birth, <37 weeks gestation, is also a risk factor for higher blood pressure in later life [20, 41, 42] and the relationship between preterm birth and blood pressure remains after adjustment for current body weight [41]. Systolic blood pressure, pulse pressure, heart rate, and diastolic blood pressure variability are elevated in young adults born preterm [43]. Furthermore, the risk for higher blood pressure increases with decreasing gestational age [44]. Total fat mass, a risk factor for cardiovascular and metabolic disease, is increased in individuals born preterm [45]. Rapid weight gain during early life significantly impacts the adverse effects of prematurity on later cardiovascular and metabolic health [46], indicating that sensitivity to early postnatal influences is heightened in preterm individuals. Rapid weight gain during infancy is also associated with higher insulin levels in young adults born premature [47], and insulin resistance is observed in adults born preterm [48]. However, preterm birth is not associated with changes in serum cholesterol or triglycerides in young adulthood [48], suggesting that differential programming is induced by slow fetal growth from preterm delivery versus IUGR at birth full term.

The blood pressure difference (<3–5 mm Hg) in adults born preterm or low birth weight is minimal [28]. However, even a small increase in blood pressure above normal across the lifespan increases the risk for cardiovascular disease in later life [49]. Thus, these studies highlight the need for further investigation into the etiology of developmental programming of hypertension and chronic disease in later life.

3.2   SEX DIFFERENCES IN BIRTH WEIGHT AND LATER CARDIOVASCULAR RISK

Men exhibit a higher blood pressure relative to age-matched women prior to menopause [50, 51]. Whether men and women differ in cardiovascular risk programmed in response to slow fetal growth is not clear. Several studies indicate that an inverse relationship is observed between birth weight and blood pressure in both men and women in young adulthood [52, 53]. Meta-analysis of 20 Nordic studies also indicates an inverse association between birth weight and blood pressure in both men and women [54]. However, few studies directly investigate whether sex differentially programs cardiovascular risk in men and women born small. One study reports a sex difference in cardiovascular risk in low birth weight children born at term [55] and young adults that is stronger in men than women [56]. Another study demonstrates that renal function is impaired to a greater degree in men born growth-restricted than women [57]. A prospective population study indicates a strong association between birth weight and blood pressure in young men that is not present in age-matched women [58]. In addition, another study reports that size at birth predicts blood pressure in women at age 60, but not at age 50 [59]. Age amplifies the association between birth weight and blood pressure in both sexes [53, 54], but whether blood pressure is altered to the same degree by age in low birth weight men and women is not entirely clear (Table 1).

A systematic review of the literature and meta-analysis indicates that the risk for higher blood pressure following premature birth is greater in women than in men [60], and gestational age demonstrates a stronger association with blood pressure than birth weight [58]. The exact mechanisms that mediate increased blood pressure in women born preterm are not clear. However, the impact of gestational age on blood pressure in women may involve alterations in the structure and functionality of the vascular tree [61]. Women born preterm also exhibit an increase in vascular resistance [62] yet, renal function remains normal [63].

Thus, whether men and women differ in the programmed response to adverse influences during fetal life or whether age impacts programming due to fetal exposure to complications during pregnancy is not clear. Additionally, the impact of sex on later cardiovascular health may differ based on whether low birth weight results from prematurity or is due to factors that slow the growth during key periods of organogenesis leading to IUGR with birth at full term.

TABLE 1: Mean systolic pressure (mmHg) at ages 59 to 71 years according to birth weight by age and sex				
	MEN		WOMEN	
BIRTH WEIGHT (POUNDS)	59–63 YEARS (n = 426)	64–79 YEARS (n = 418)	60–63 YEARS (n = 203)	64–71 YEARS (n = 184)
To 5.5	166 (n = 17)	171 (n = 18)	156 (n = 10)	169 (n = 9)
To 6.5	163 (n = 53)	168 (n = 55)	156 (n = 35)	165 (n = 33)
To 7.5	162 (n = 120)	1168 (n = 144)	162 (n = 75)	160 (n = 68)
To 8.5	163 (n = 136)	165 (n = 111)	163 (n = 48)	163 (n = 48)
>8.5	161 (n = 100)	163 (n = 92)	152 (n = 137)	155 (n = 26)
Total	162	166	159	161
Standard deviation	23	24	23	26

3.3   BIRTH WEIGHT AND ENHANCED SUSCEPTIBILITY

Several studies suggest that slow fetal growth increases susceptibility to secondary renal disease. Low birth weight enhances the risk of chronic renal failure in individuals with hypertension and/or diabetes relative to normal birth weight counterparts [64]. Onset of end stage renal disease is earlier in low birth weight individuals with autosomal dominant polycystic kidney disease, a genetic disease that involves a loss of renal function following the development of renal cysts [65]. IUGR adversely impacts children with nephrotic syndrome relative to normal birth weight counterparts [66]. Thus, low birth weight is associated with an enhanced susceptibility to renal injury and disease. The mechanisms involved are not clear, but may involve impaired renal development programmed by

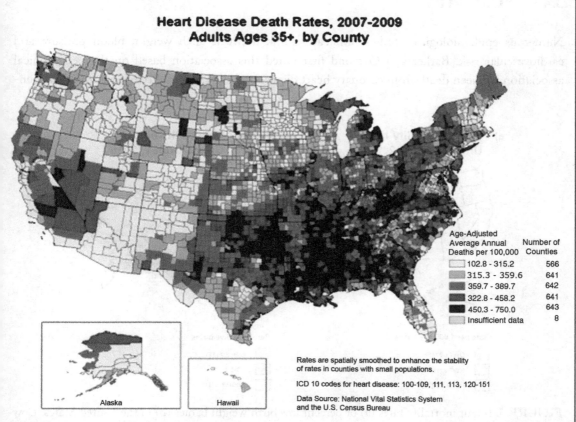

Heart Disease Death Rates, 2007-2009
Adults Ages 35+, by County

Age-Adjusted
Average Annual   Number of
Deaths per 100,000   Counties

	Age-Adjusted Average Annual Deaths per 100,000	Number of Counties
	102.8 - 315.2	566
	315.3 - 359.6	641
	359.7 - 389.7	642
	322.8 - 458.2	641
	450.3 - 750.0	643
	Insufficient data	8

Rates are spatially smoothed to enhance the stability
of rates in counties with small populations.

ICD 10 codes for heart disease: 100-109, 111, 113, 120-151

Data Source: National Vital Statistics System
and the U.S. Census Bureau

Alaska   Hawaii

FIGURE 4: Heart disease death rates from 2007 to 2009 in adults 35 years or older by US County. Used with permission from the Centers for Disease Control, Data Source, National Vital Statistics System.

poor fetal growth that alters the structure and physiology of the kidney, leading to enhanced vulnerability to renal injury and disease.

Pregnancy serves as a natural stress test for future cardiovascular disease [67]. Risk of a preeclamptic pregnancy is elevated in low birth weight women or women born preterm [68, 69]. Low birth weight is also associated with an increased risk for gestational diabetes [70, 71] or preterm delivery [72]. Low birth weight shares many of the common risk factors noted for preeclampsia or gestational diabetes including endothelial dysfunction and increased BMI [36, 73]. Yet, whether the increased risk of a complicated pregnancy in low birth weight women is due to the presence of risk factors related to chronic disease, or if it is a direct result from their exposure to slow growth during their fetal life is not clear. However, these findings highlight the impact that a complicated pregnancy in one generation has on the chronic health of subsequent generations.

3.4   SUMMARY

Numerous epidemiological studies implicate a link between birth weight, blood pressure and cardiovascular risk. Barker and Osmond first noted this association based on the geographical association between death from coronary heart disease and infant mortality rates in the same gen-

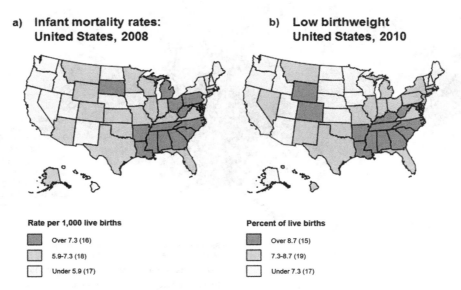

FIGURE 5: Infant mortality rates (a) or percent low birth weight babies (b) in the United States. Low birth weight is less than 2500 g (5.5 pounds). Source National Center for Health Statistics, final natality data. Used with permission from the March of Dimes Peristats.

eration [6]. This same geographical association is evident when examining the geographical distribution of death rates from heart disease (Figure 4) with infant mortality rates (Figure 5a) or percent of low birth weight (Figure 5b) from recent rankings in the United States. However, the ability to investigate direct cause and effect within the human population is limited. Thus, use of experimental models that mimic the causes of low birth weight can be utilized to provide insight into how adverse influences during fetal and early life program an increased risk for hypertension and cardiovascular and renal disease in later life.

. . . .

CHAPTER 4
Experimental Models of Developmental Programming

Animal models that mimic the human condition of slow fetal growth provide proof of principle for the theory of developmental origins of health and disease. The causes of low birth weight are numerous and include prematurity or slow fetal growth due to maternal diseases and infections, smoking, alcohol and drugs, placental problems, poor nutrition, stress, and fetal infection. Different methods have been utilized to induce a suboptimal environment during development. Despite the manipulation utilized, a common phenotypic outcome such as hypertension or high blood pressure is observed in these models. Importantly, insight from these different models of developmental insult indicates that similar mechanistic pathways contribute to the etiology of programmed hypertension and cardiovascular disease.

4.1   ANIMAL SPECIES

Species utilized for mechanistic investigation into the developmental origins of adult health and disease are numerous and include the mouse, rat, guinea pig, sheep, pig, and monkey. Similar methods are used to induce programming of later cardiovascular risk in these different species and include maternal undernutrition [74, 75], exposure to maternal glucocorticoids [76] or hypoxia [77, 78], and placental insufficiency [79]. Rodents are widely used primarily due to their inexpensive cost and their short gestational period and lifespan. Their short reproductive cycle also facilitates the study of multiple generations. Most rodent species are inbred removing the influence of genetic differences so that the researcher can focus on environmental effects such as stress or nutrition on the programming of later chronic disease. Use of the mouse also provides access to an extensive number of transgenic animals that can be utilized to investigate the relative importance of specific genes through the use of conditional or tissue-specific knockouts. Mice have many genes and molecular pathways in common with humans, yet may not be the best species for preclinical studies as drugs that work well in mice may not always be effective in humans. In addition, renal development in rodents extends beyond birth thus differing from nephrogenesis in the human which is completed prior to birth [80, 81]. Larger species are useful as they provide gestational lengths similar to that of the human. Additionally, both the mother and fetus can be studied during development making them valuable to the field of developmental programming. Yet, cost for care and maintenance of the larger species in addition to ethical considerations can impact feasibility for their use in experimental studies.

Despite the species or the method of developmental insult, common risk factors for cardiovascular disease are observed in models of developmental programming of chronic health. Hypertension is programmed by developmental insult in the mouse [82], the rat [75], the guinea pig [83], and the sheep [84]. A reduction in nephron number is observed in sheep and rats exposed to maternal undernutrition [75, 84] and placental insufficiency [85, 86]. Fetal exposure to maternal glucocorticoids also programs a reduction in nephron number in the mouse [87], rat [88] and sheep [89]. Vascular dysfunction is a common consequence of developmental insult and impaired vascular function is observed following undernutrition during fetal life in the mouse [90], the rat [91], and sheep [92]. Thus, common phenotypic outcomes observed following different methods of developmental insult and in different animal species implicates common pathways in the etiology of developmental programming of chronic disease. These findings highlight the relevance of these models for use in the mechanistic investigation into the etiology of cardiovascular risk programmed by a developmental insult.

4.2   TIMING OF THE INSULT

The kidney is known to play a key role in the long-term control of blood pressure [93]. In humans, nephrogenesis is complete by late gestation [80]; however, kidney development in the rat proceeds until after birth [81] (Table 2). Thus, the period vulnerable to the impact of programming in the rat includes the period of fetal and postnatal life that coincides with nephrogenesis. The importance of timing of developmental insult is demonstrated in rodent models whereby exposure to adverse stimuli can program later hypertension regardless of whether the insult coincides with nephrogenesis during fetal [74, 76, 79] or postnatal life [94, 95]. Thus, experimental studies demonstrate that susceptibility to later chronic health results from adverse influences that occur during key periods of development.

4.3   DIETARY RESTRICTION

Fetal nutrition is a key factor in the control of fetal growth [96]. One of the first models used to investigate the hypothesis of fetal programming proposed by David Barker involved manipulation of the maternal diet during gestation in the rat [74]. Typical protein content in rodent chow is between

TABLE 2: Approximate period over which nephrogenesis occurs in a number of species.

SPECIES	LENGTH OF PREGNANCY	APPROXIMATE PERIOD OF NEPHROGENESIS	RELATIVE PERIOD OF GESTATION
Human	40 weeks	5–36 weeks	12–90%
Sheep	150 days	30–130 days	20–90%
Guinea pig	63 days	22–55 days	35–90%
Spiny mouse	40 days	19–37 days	50–90%
Mouse	20 days	E11-PN5-7 days	55–125%
Rat	22 days	E12-PN8-10	55–140%
Rabbit	32 days	E12-PN21	35–160%
Pig	112 days	E20-PN21-25	20–120%

Note: The first 4 species complete nephrogenesis prior to birth while the remaining species complete nephrogenesis in the postnatal period. E = embryonic day and PN = postnatal day.

18% and 20%. Maternal diets composed of moderate (9% protein) to severe (5% protein) protein composition during gestation and lactation are used by many investigators to study the impact of poor nutrition on fetal growth and later cardiovascular risk in the offspring [74, 75, 97]. Moderate to severe reductions in protein content during gestation and lactation programs growth-restricted pups that develop an increase in blood pressure in later life relative to pups born to pregnant rats maintained on regular chow [74]. Cardiac dysfunction is also observed in offspring exposed to maternal undernutrition during gestational life [98]. In addition, fetal exposure to moderate maternal global undernutrition also programs an increase in blood pressure in offspring, an effect observed in multiple species including the rat [99], the guinea pig [83], and the sheep [100].

4.4   REDUCED UTEROPLACENTAL PERFUSION

Fetal growth can also be impaired by reductions in uteroplacental perfusion to the fetus [101]. Placental insufficiency is the major cause of intrauterine growth restriction [9], and numerous investigators utilize different methods of reduced uterine perfusion to study the developmental origins of chronic disease. Rodent models of reduced uteroplacental perfusion vary by method and include

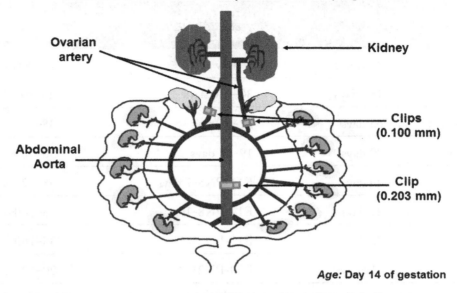

Model of intrauterine growth restriction (IUGR):
Induction of reduced uterine perfusion in the pregnant rat.

FIGURE 6: The reduced uterine perfusion model of placental insufficiency induced at day 14 of gestation in the rat. Used with permission from *Methods in Molecular Medicine* 2005;122 383–392.

bilateral ligation of the uterine artery and vein [85] or an overall reduction in uteroplacental flow induced by placement of silver clips with set internal diameters on the ovarian arteries and the abdominal artery [79] (Figure 6). Timing of reduced uterine perfusion in the rat can vary from day 14 of gestation [79] to day 18 of gestation [85] during the total 22-day gestational period in the rat. However, regardless of the method used to reduce uteroplacental perfusion in the rat, a marked increase in blood pressure is observed in male growth-restricted offspring in adult life [79, 85]. Models of placental insufficiency in the sheep also vary by method and involve embolization of the placental vascular bed [102] or exposure to an elevated ambient temperature which increases vascular resistance in blood flow the placenta and fetus leading to IUGR and programming of cardiovascular risk in the offspring [103].

4.5   HYPOXIA

Hypoxia or a reduction in oxygen delivery to the fetus is a common adverse complication that can occur during pregnancy [104]. Intrauterine growth restriction induced by fetal exposure to maternal hypoxia in experimental studies programs an increase in cardiovascular risk as indicated by adverse cardiac remodeling [105, 106] and impaired vascular function [77, 78], observations noted in the rat

[77, 105, 106] and the sheep [78]. Maternal hypoxia also programs a marked reduction in nephron number and a marked increase in blood pressure that occurs with age in the rat [107].

4.6   PHARMACOLOGICAL

Numerous pharmacological agents that block factor keys to proper development are utilized to program hypertension and increased cardiovascular risk in experimental studies. Glucocorticoids are a class of steroid hormones that bind to the glucocorticoid receptor and regulate a number of physiological processes including immune, cardiovascular, metabolic, and homeostatic [108]. Glucocorticoids also play a key role in development and are implicated in fetal growth restriction [109]. Prenatal exposure to glucocorticoids in the rat programs hypertension in male offspring [76], demonstrating that hypertension in adult life can be programmed by prenatal exposure to maternal glucocorticoids. The renin angiotensin system (RAS) is a key regulator in the long-term control of body fluid homeostasis and arterial pressure [110]. However, the RAS is also critical for proper renal development [111]. Inhibition of the RAS during the first 10 days of life in the rat coinciding with the final stages of nephrogenesis programs a marked increase in blood pressure in later life [94]. Cyclooxygenase-2 (COX-2) also plays an important role in renal development [112]. Inhibition of COX-2 during nephrogenesis in the rat also programs hypertension associated with a reduction in nephron number and renal injury [113]. Thus, findings from these studies utilizing pharmacological blockade during early postnatal life highlight that vulnerability to programmed cardiovascular risk in the rat extends after birth and through the final stages of renal development.

4.7   NICOTINE

Cigarette smoke is a significant health risk for the fetus during pregnancy and is associated with a significant risk for low birth weight and increased blood pressure in offspring after birth [114]. Prenatal exposure to nicotine in the rat enhances vascular sensitivity and augments the blood pressure response to angiotensin II, a potent vasoconstrictor [115]. Thus, offspring exposed to nicotine during prenatal life have a greater vasoconstrictor response within the vasculature and a greater increase in blood pressure when exposed to angiotensin II compared to offspring that were not exposed to nicotine during prenatal life. Therefore, these studies indicate that prenatal exposure to cigarette smoke or nicotine programs increased cardiovascular risk in later life.

4.8   EARLY LIFE STRESS

Early life stress is highly associated with an increase in cardiovascular risk in the human population [116]. Whether this observation is correlative or directly linked is demonstrated by experimental models that utilize maternal separation to induce chronic early life stress. Maternal separation involves separation of the pups during early postnatal life from the mother for several hours per day [117]. Rat offspring in this model of early life stress demonstrate enhanced vascular and blood pressure sensitivity to angiotensin II indicating that stress in early life may predict a greater susceptibility to cardiovascular disease in later life [117, 118].

4.9   MATERNAL OBESITY

Early investigation into the mechanisms of developmental programming of adult disease focused on studies related to the impact of undernutrition in early life. Experimental studies utilizing undernutrition during fetal life provided cause and effect for the observation that birth weight was inversely proportional to blood pressure and death from coronary heart disease. However, obesity is a major health risk in the Western world and obesity during pregnancy is associated with a host of complications including hypertension and gestational diabetes [119]. Maternal obesity is also associated with significant long-term risk for chronic disease in the offspring including obesity, type 2 diabetes, and hypertension [39, 120]. Rodent models of maternal obesity induced by lard [121] or energy rich diets [122] are being utilized to investigate the impact of maternal obesity on later cardiovascular and metabolic [123] health of the offspring.

4.10   MATERNAL DIABETES

Maternal diabetes leads to raised blood pressure in children [124] in a manner that may be sex-specific [125]. Experimental models utilize numerous methods to mimic the human condition of diabetes including injection of glucose and/or streptozotocin (STZ), a pharmacological agent that selectively destroys beta-cells, with some degree of blood glucose maintained by insulin [126, 127]. Blood pressure in rodent offspring exposed to different experimental models of maternal diabetes is elevated [126] and can also result in sex-specific increases in blood pressure in the offspring [127].

4.11   EARLY ACCELERATED GROWTH

Early postnatal growth in the human can be significantly altered by dietary manipulation during the first nine months of life [128]. Early accelerated growth in the human is an independent risk factor for chronic disease [129] and accelerated growth in early life adversely affects blood pressure in later life [42]. Postnatal overfeeding induces hypertension and renal dysfunction regardless of birth weight in the rat [130]. Thus, despite differences in the timing of nephrogenesis in the human and the rat, the developmental period sensitive to the programming effects of an adverse exposure includes early postnatal life. Methods to induce early accelerated growth in the rat include reduction of litter size at birth [130] or exposure to a maternal high fat diet during lactation [131].

4.12 IMPACT OF SEX

Many experimental models of developmental programming of hypertension exhibit a sex difference in the programmed phenotypic outcome or response to a developmental insult. Female offspring are more resistant to the programming of hypertension induced by exposure to moderate maternal undernutrition including moderate maternal protein restriction (9% versus 18–20% protein content in the maternal diet) [91, 132], moderate maternal global nutrient restriction [99] or uteroplacental insufficiency [79, 133]. Thus, male offspring exhibit a marked increase in blood pressure in adulthood compared to their same-sex control counterparts whereas female offspring remain normotensive [79, 99, 132, 133] (Figure 7). Other developmental insults in the rat including prenatal

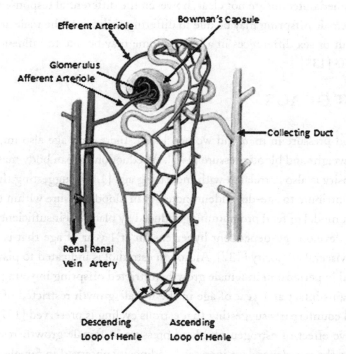

FIGURE 7: Measure of mean arterial blood pressure in a rat model of intrauterine growth restriction (IUGR) induced by reduced uterine perfusion. Data shown are for both male and female IUGR versus control offspring, at 4, 8, and 12 weeks of age. *$P < 0.05$ vs. male control; †$P < 0.05$ vs. female control; ‡$P < 0.01$ vs. control; and §$P < 0.01$ vs. control. All data are mean ± SEM. Used with permission from *Hypertension* 2003;41: 457–462.

exposure to nicotine [113], postnatal exposure to maternal separation or stress [117], postnatal inhibition of COX-2 [113], and early postnatal hypernutrition [130] program a sex difference in adult cardiovascular risk with male offspring more sensitive to the specific developmental insult relative to their female littermates Sex differences are also observed in other species besides the rat. Maternal administration of glucocorticoids during gestation in the sheep programs a sex difference in adult renal function [134]. However, sex differences in models of fetal undernutrition in the rodent can be abolished by exposure to a greater severity of fetal insult (5% protein content in the maternal diet versus 18–20% protein content) resulting in hypertension associated with a reduction in nephron number in both male and female offspring [135]. Yet, in rodent models of maternal obesity, female offspring are more sensitive to the maternal insult and exhibit an increase in blood pressure that is not observed in male littermates [121]. Fetal exposure to maternal diabetes in the rat programs hypertension in male, but not female offspring [127]. Thus, sex differences in programming of cardiovascular risk are insult-specific. In addition, sex differences in blood pressure in many models of developmental insult may be impacted with age [136, 137]. The mechanisms that mediate sex differences in programmed outcome are not clear; however, the differential response to developmental insult in male and female offspring may be due to differing abilities of the male and female fetus to adapt to a fetal insult or sex differences in adult outcome may be due to influences exerted by sex hormones in later life [138].

4.13   IMPACT OF AGE

Aging impacts blood pressure in men and women [139]. Increasing age also impacts the association between birth weight and blood pressure [140]. A reduction in lean body mass associated with an increase in adiposity is also correlated with increasing age [141], suggesting that obesity related mechanisms may contribute to age-dependent increases in blood pressure within the general population. In the rodent model of fetal programming induced by placental insufficiency female growth-restricted offspring develop age-dependent hypertension at 1 year of age that is associated with a marked increase in visceral adiposity [137]. Although estradiol is indicated to play a protective role against programmed hypertension in female growth-restricted offspring in young adulthood [142], uterine weight remains intact at 1 year of age in the female growth restricted offspring relative to their female control counterparts suggesting that estrous cycling is preserved [137]. Whether aging lessens the protective effect of estrogen on blood pressure in female growth-restricted rats is not yet known; however, the age-dependent increase in adiposity observed in female growth-restricted offspring may serve as an underlying mechanism. In the model induced by postnatal blockade of the RAS, programming of hypertension and a marked reduction in nephron number are associated with an age-dependent increase in proteinuria, a marker for renal injury [143], and angiotensin II-dependent hypertension [144] that is sex-dependent. Thus, age induces a greater susceptibility to cardiovascular/renal risk in many experimental models of developmental insult; however, the mechanisms are not yet clear.

4.14   SUMMARY

Pregnancy complications affect approximately 10% of all pregnancies. Not only do complications during pregnancy compromise maternal and fetal health, but they may also program hypertension and increased cardiovascular risk in the offspring. Complications that can compromise the future health of the offspring are numerous and include insults that result from a reduction in nutrient delivery to the fetus, involve exposure of the fetus to maternal hypoxia, glucocorticoids or stress, impaired glucose homeostasis or the adverse environment milieu of maternal obesity. Long-term consequences are similar despite the species or method of developmental insult making them valuable tools for investigation into cause and effect and in-depth study of the mechanisms by which adverse events during critical periods of development program hypertension and cardiovascular disease in the offspring.

·   ·   ·   ·

CHAPTER 5

Mechanisms of Developmental Programming

Experimental models of developmental programming of hypertension are providing significant insight in the mechanisms by which an insult in early life programs increased cardiovascular risk. Despite differences in the type of insult, timing of insult (prenatal versus postnatal), and/or the type of species utilized, experimental models are demonstrating that early life responses to an adverse event or environmental exposure program hypertension and altered cardiovascular health through common mechanistic pathways.

5.1   NEPHRON NUMBER

The kidneys play a key role in the long-term control of blood pressure regulation through the control (excretion or retention) of sodium and water by the kidney [93]. The functional unit of the kidney is the nephron which is composed of a glomerulus that filters fluid, electrolytes including sodium and chloride, nutrients and waste products, and a tubule where fluid is converted into urine (Figure 8). In humans, nephron number is established during fetal life [145]. Illness, injury and aging can reduce nephron number after birth, but nephron number cannot be increased or restored after birth [145]. Low birth weight in humans is associated with a marked reduction in nephron number [145] (Figure 9). A reduction in nephron number is also observed in many models of developmental programming induced by placental insufficiency [146], maternal protein restriction [97, 135], prenatal exposure to glucocorticoids [88], and pharmacological blockade induced by inhibition of the RAS [94] or COX-2 [113]. Thus, the events that program low birth weight also alter renal development resulting in a reduction in nephron number.

Hypertension is associated with a reduction in nephron number in many experimental models of developmental programming [75, 88, 97], suggesting that a loss of nephrons during development may serve as the link between birth weight and blood pressure [145, 146]. Whether a reduction in nephron number is the cause of the programmed hypertension, however, is not clear. Long-term cardiovascular and renal consequences following unilateral nephrectomy (removal of

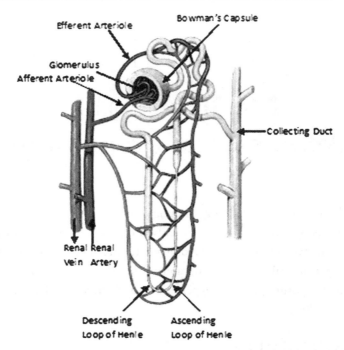

FIGURE 8: Basic components of the nephron. Used with permission from the *Kidney Study Guide, Biomedweekly, 2013.*

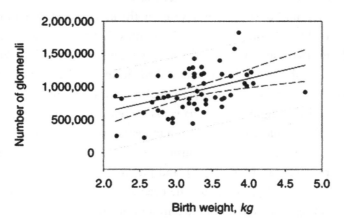

FIGURE 9: The relationship between birth weight and total glomerular number among all cases that includes infants, children, and adults. Symbols are (•) N_{glom} vs. birth weight; (——) N_{glom} vs. birth weight regression; (—) 95% regression CI; (......) regression prediction interval. The regression coefficient predicts a gain of 257,426 glomeruli per kg increase in birth weight, $r = 0.423$, $P = 0.0012$, $N = 56$. Used with permission from *Kidney Int.* 2003;63: 2113–2122 (Figure 4).

one kidney) in the adult human are similar to that observed in the general population [147]. Thus, removal of a kidney which is equivalent to a 50% reduction in nephron number after development or in adulthood does not result in a significant impairment in blood pressure regulation. However, experimental studies indicate that uninephrectomy during renal development is associated with later hypertension [148, 149]. Further evidence linking a reduction in nephron number that occurs during development to hypertension is demonstrated by genetic models that result in a naturally occurring deficient in nephron number [150] and mice that exhibit a reduction in nephron number induced by prematurity [151]. However, prenatal exposure to glucocorticoids does not always result in a reduction in nephron number in the hypertensive offspring [88] indicating that a reduction in nephron number per se is not an absolute requirement for the fetal programming of hypertension. A reduction in nephron number following developmental insult is often associated with compensatory glomerular hypertrophy in order to sustain adequate renal function [75]. This may accelerate the loss of functional nephrons over time leading to a reduction in the filtration surface area impacting the kidneys' ability to maintain fluid and sodium homeostasis and blood pressure. To conclude, these experimental studies indicate that a significant reduction in nephron number that occurs during nephrogenesis may be associated with hypertension in later life. Although a reduction in nephron number may not be the direct cause of the increase in blood pressure observed in response to a developmental insult, a reduction in nephron number that occurs during development may contribute to the increased risk of cardiovascular and renal disease that can occur with aging.

5.2   RENAL FUNCTION

Glomerular filtration rate (GFR) is a method used to determine how well the kidneys are working. It represents a measure of how well a substance is cleared from the blood and excreted by the kidneys [152]. Whether renal function is reduced or increased in low birth weight and individuals born preterm (as reviewed in 146) or in experimental models of developmental programming [79, 135] is not clear. However, a reduction in nephron number may increase the risk for the development of chronic kidney disease and impaired renal function in low birth weight individuals in later life [153]. Moreover, other factors such as aging, hypertension, diabetes, and obesity may exacerbate this risk [153].

5.3   SEX HORMONES

Testosterone is the main male sex hormone. Testosterone levels are significantly elevated in male growth restricted rats relative to their male control counterparts in the model of fetal programming induced by placental insufficiency in the rat [154]. Castration or removal of the male sex hormones in this model abolishes hypertension in adult male growth restricted rats [154] suggesting that hypertension in male growth-restricted offspring in this model is testosterone-dependent. Yet, the sex difference in adult blood pressure in this model of IUGR may not be related only to the presence of increased testosterone in the adult male growth restricted rat. Female growth restricted offspring in this model of IUGR induced by placental insufficiency are normotensive in adulthood [79]. Circulating levels of estradiol, the most important form of estrogen, do not differ in female growth restricted rats relative to age-matched, same-sex controls. However, ovariectomy or removal of the ovarian hormones induces hypertension in young adult female growth restricted rats with no effect on blood pressure in age-matched female control rats [142]. Importantly, ovariectomy-induced hypertension is reversed by treatment with estradiol [142]. Thus, these findings indicate that the sex hormones contribute to the regulation of adult pressure in this model of IUGR, whereby testosterone contributes to the development of hypertension in male growth-restricted rats and estrogen is protective against the development of hypertension in female growth-restricted rats in young adulthood. Other models of developmental programming also demonstrate a sex difference that may be mediated by the influences of sex hormones. Enhanced sensitivity to angiotensin II, a potent vasoconstrictor, is associated with a marked increase in testosterone in male offspring exposed to early life stress induced by maternal separation [117]. Castration abolishes the enhanced vasoconstrictor response to acute angiotensin II in male offspring following early life stress eliminating the sex difference in cardiovascular risk [117]. Although the importance of the female sex hormones has not yet been investigated in this model of early life stress, these findings suggest that testosterone contributes to the programming of increased cardiovascular risk in this model. Yet, a key role for the male sex hormones is not indicated in all experimental models that demonstrate a sex difference in adult blood pressure. Castration does not abolish hypertension induced by prenatal exposure to low protein in male rats [155]. Ovariectomy, however, induces hypertension at a younger age in female rats exposed to maternal low protein [136]. The exact mechanisms by which sex hormones contribute to hypertension in these different models of developmental programming of hypertension and adult chronic disease have not been clearly elucidated. However, the etiology of programmed cardiovascular risk may involve modulation of regulatory systems that control sodium and fluid balance and the long-term regulation of blood pressure.

5.4   RENIN ANGIOTENSIN SYSTEM

5.4.1   Developmental Programming of the RAS and Hypertension

Angiotensin II is a potent vasoconstrictor that mediates its actions via its receptor, the angiotensin type 1 receptor (AT1R). The angiotensin type 2 receptor (AT2R) plays a key role in development and works to oppose the vasoconstrictor actions of the AT1R [156]. Systemic blockade of the RAS abolishes hypertension induced by gestational exposure to low protein [157], global maternal nutrient restriction [158], and placental insufficiency in the rat [154], suggesting a key role for the RAS in the etiology of hypertension programmed in response to a fetal insult. Alterations in expression of components of the RAS are observed in many models of programmed hypertension, including those induced by fetal exposure to maternal diabetes [159], maternal protein restriction [160], maternal glucocorticoids [161], and placental insufficiency [162]. Moreover, programming of the RAS is sex-specific in many of these models, and numerous studies indicate that sex-specific programming of the RAS may contribute to the sexual dimorphism of blood pressure programmed by a developmental insult.

5.4.2   Sex Differences in the Developmental Programming of the Renal RAS

Components of the renin angiotensin system (RAS) are expressed throughout the kidney and play an important role in the long-term regulation of blood pressure [163] (Figure 10). Angiotensin-converting enzyme-2 (ACE2), angiotensin-(-1-7) (Ang-(1-7), and the mass receptor represent an endogenous pathway that is counter-regulatory to the action of ACE, angiotensin II and the AT1R [164, 165]. Renal AT2R and ACE2 expressions are elevated in female rats relative to males [166]. Moreover, renal AT2R expression can be up-regulated by estrogen [167] and estrogen is reported to

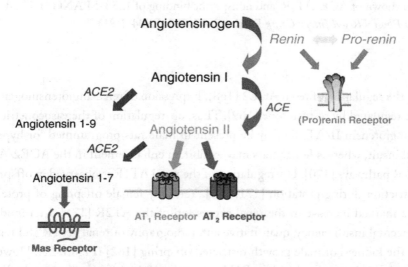

FIGURE 10: Schematic of the renin-angiotensin system. Used with permission from *Experimental Cell Research Volume 318, Issue 9 2012 1049–1056* (Figure 1).

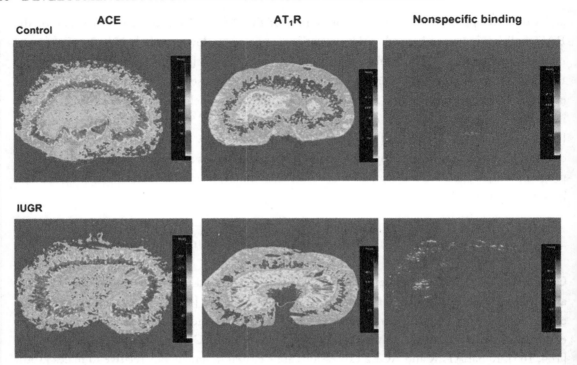

FIGURE 11: Quantitative autoradiography of angiotensin type 1 (AT$_1$R) receptors and ACE in whole kidney from male control and male growth restricted offspring at 16 weeks of age. Representative auto-radiographs are shown of ACE, AT$_1$R, and nonspecific binding of 125 I-SI ANG II. Used with permission from *Am J Physiol Regul Integr Comp Physiol.* 2007; 293: R804–R811.

contributes to the regulation of renal ACE2 [168]. Expression of renal angiotensinogen, however, is reported to be regulated by testosterone [169]. Thus, up-regulation of the vasoconstrictor pathway of the RAS (angiotensin II/AT1R) may be present in male rats programmed for hypertension by developmental insult, whereas female rats may exhibit an enhancement in the ACE2/Angiotensin-(1-7) and AT2R pathways [170]. Up-regulation of the renal AT1R is observed in offspring exposed to protein restriction during gestation [171, 160]. Yet, only female offspring of protein restricted dams exhibit a marked increase in the renal expression of the AT2R [171]. In a model of IUGR induced by placental insufficiency, quantitative autoradiography of renal AT1R and renal ACE are not altered in the kidneys of male growth restricted offspring [162] (Figure 11). However, expression of renal angiotensinogen and renin mRNA expression, initial components in the RAS pathway are up-regulated in male growth restricted offspring that exhibit hypertension in adulthood [162].

Renal expression of ACE2, however, is elevated in their normotensive female growth restricted rats counterparts [142]. A marked reduction in renal ACE2 is associated with induction of hypertension by ovariectomy in the female growth restricted rat [142], indicating that loss of the protective effects of the ACE2/Ang-(1-7) pathway may contribute to hypertension induced by ovariectomy in this model of sex-specific fetal programming of hypertension. Thus, sex-specific modulation of the RAS by sex steroids may contribute to sex differences in adult blood pressure programmed by fetal insult.

5.4.3   Developmental Programming of the Central RAS

Central actions of the RAS influence cardiovascular function and blood pressure [172]. Sheep exposed to glucocorticoids during fetal life exhibit an increase in expression of the AT1R in the hypothalamus, an area of the brain involved in the regulation of fluid balance and blood pressure [173]. Maternal undernutrition in the mouse programs an increase in mRNA expression of ACE and a reduction in AT2R in the fetal brain [174]. Inhibition of the renin angiotensin system via AT1R blockade or blockade of the angiotensin converting enzyme administered intracerebroventricularly significant reduces blood pressure in offspring programmed for hypertension via prenatal exposure to maternal protein restriction [175]. Thus, these studies suggest that alterations in the central RAS contribute to the developmental programming of hypertension in addition to contributions from the peripheral and renal RAS.

5.4.4   Developmental Programming of Enhanced Angiotensin II Sensitivity

Numerous models of developmental insult exhibit enhanced sensitivity to angiotensin II [115, 117, 176]. Sex-specific responses are noted with males more sensitive than females [115, 117, 176, 177]. Castration abolishes enhanced sensitivity to acute angiotensin II in male growth-restricted rats programmed for hypertension via fetal exposure to placental insufficiency [176] and in male rats exposed to early life stress [118]. Thus, these studies indicate that modulation of the RAS via testosterone contributes to the etiology of hypertension induced by developmental insult in the male rat.

5.4.5   The RAS and the Developmental Programming of Nephron Number

Temporal changes in the RAS are observed in experimental models of developmental programming of hypertension. Renal renin and angiotensinogen expressions are elevated in adult male growth restricted rats programmed by fetal exposure to placental insufficiency; yet expression of renin and angiotensinogen are significantly decreased in the kidney at birth [162]. Blockade of the RAS during early life coinciding with nephrogenesis programs a marked reduction in nephron number linked to hypertension in the adult animal [94], indicating a key importance of the RAS in nephrogenesis. Renin expression is also reduced at birth in offspring exposed to maternal low protein [75] and a reduction in nephron number associated with hypertension in the adult low protein offspring. Thus, these studies suggest that fetal insults that reduce expression of the RAS program a reduction in nephron number that may impact chronic health in later life.

5.5 SYMPATHETIC NERVOUS SYSTEM AND THE RENAL NERVES

The kidneys regulate the long-term control of blood pressure via their ability to maintain body fluid and electrolyte balance [93]. However, activation of the sympathetic nervous system (SNS) plays a key role in hypertension with an increase in renal sympathetic nerve activity (RSNA) serving as a link between the central SNS and the kidney [178]. Whether the SNS activity is elevated in low birth weight individuals is controversial [179, 1780]. Yet, experimental studies demonstrate that the SNS contributes to hypertension induced by exposure to maternal obesity [122], and that the renal nerves contribute to hypertension in male growth restricted offspring exposed to placental insufficiency [137,181] (Figure 12) or glucocorticoids during fetal life [182]. Moreover, the mechanism by which renal denervation abolishes hypertension programmed by prenatal insult may involve reduced abundance of renal sodium transporters within the proximal tubule of the nephron [182].

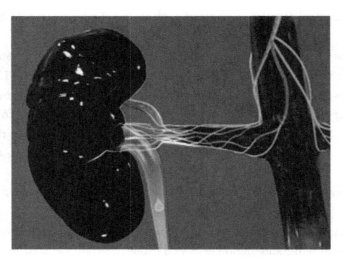

FIGURE 12: Diagram of the renal nerves. Used with permission from *Meditronics.*

5.6   RENAL SODIUM TRANSPORT

Sodium reabsorption within the kidney is regulated by sodium transporters and sodium balance or homeostasis plays a critical role in the regulation of blood pressure [183]. Increased sodium transport is one potential mechanism linking prenatal insult with later adult hypertension. A significant increase in renal sodium transporter abundance [184, 185] associated with an increase in proximal tubule transport [186] is noted in several models of developmental programming of hypertension induced by undernutrition during fetal life. Therefore, programming in response to developmental insult may alter the kidney's ability to regulate sodium balance via alterations in sodium transport within the renal tubule.

5.7   VASCULAR FUNCTION

Vascular health predicts later cardiovascular risk. Impaired vascular health includes changes in arterial stiffness and the ability of the vasculature to respond properly to vasoactive factors, or factors that can dilate or constrict the blood vessel [187]. Arterial stiffness, a marker of increased cardiovascular risk, is elevated low birth weight children, and in offspring exposed to an experimental model of placental insufficiency [103]. Aortic intima-media thickness, a noninvasive marker of preclinical vascular disease, is increased at 18 months of age in children born with IUGR [188]. Similar findings are observed in experimental models of cardiovascular programming induced by prenatal exposure to nicotine [189]. Altered vascular structure is also observed in young adults born preterm [61] or individuals born IUGR but with birth at full term [190], suggesting that programming of vascular function may be independent of gestational age. Children exposed to a diabetic pregnancy have an increase in markers of endothelial activation [21] indicative of increased cardiovascular risk. Slow fetal growth is also associated with higher systolic blood pressure and impaired flow-mediated dilatation in young adults [35]. Hypertension programmed by prenatal exposure to placental insufficiency in the rat is associated with impaired vascular function [191]. Vascular dysfunction is also observed in rodent offspring of a diabetic pregnancy [192]. Prenatal exposure to global maternal nutrient restriction programs hypertension and impaired vascular function in a manner that is sex- and age-dependent [99]. Maternal obesity programs vascular dysfunction in both sexes; yet only female offspring exhibit a marked increase in blood pressure [121]. Age impacts vascular function in offspring programmed by prenatal exposure to a hypoxic environment [193]. Additionally, programming of vascular dysfunction can be transmitted to the next generation in the absence of additional stressor indicating that programming of cardiovascular risk in one generation may be transmitted to the next generation [194]. Thus, these studies indicate that programming of impaired vascular health is observed in humans and experimental models of developmental insult. Additionally, these studies demonstrate that vascular dysfunction persists into adulthood, can be further impacted by age and sex, and that programming of impaired vascular function can be transgenerational. Importantly, programmed disturbances in vascular health and hypertension can be abolished by prenatal micronutrient or antioxidant interventions [195] indicating that undernutrition is a direct mediator of developmental programming of later cardiovascular risk and that developmental exposure to oxidative stress may serve as a contributory factor.

5.8   OXIDATIVE STRESS

5.8.1   Developmental Programming, Oxidative Stress, and Hypertension

Oxidative stress plays a key role in cardiovascular diseases including hypertension. Oxidative stress results from an increased generation of reactive oxygen species and/or a diminished antioxidant capacity [196] (Figure 13). Antioxidant capacity is altered by poor fetal growth predisposing low birth weight babies to increased oxidative stress [197]. Antenatal exposure to antioxidants can prevent hypertension and vascular function programmed by prenatal exposure to maternal protein restriction in the rat [198], highlighting the importance of oxidant status during development on fetal growth and later chronic health. Markers of oxidative stress are elevated in children with low birth weight [199] indicating that alterations in oxidant status may persist following slow fetal growth. Experimental models of developmental programming also exhibit an increase in oxidative stress in adult offspring that exhibit hypertension [200, 201, 202] or cardiovascular dysfunction [106] in later life. Treatment with antioxidants reduces blood pressure in experimental models programmed by placental insufficiency [200], gestational protein restriction [201], and global nutrition restriction during fetal life in the rat [202], indicating a key role for oxidative stress in the developmental programming of hypertension. Moreover, treatment of hypoxic dams with an antioxidant during gestation ameliorates vascular dysfunction in the offspring [106], indicating antioxidant therapy as a potential interventional target against programming of chronic disease.

5.8.2   Sex Differences in the Developmental Programming of the Oxidative Stress

Oxidative stress also contributes to sex differences in adult blood pressure programmed by fetal insult. Placental insufficiency programs a marked increase in markers of oxidative stress associated with hypertension in the male growth restricted rat that is significantly lowered by antioxidant treatment [200]. However, female growth-restricted offspring do not exhibit an increase in markers of oxidative stress, they are normotensive in adulthood relative to their male littermates, and

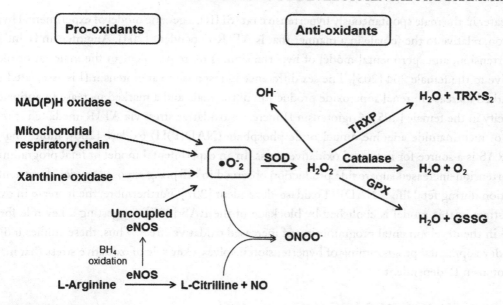

FIGURE 13: Possible role of reactive oxygen species in vascular remodeling and implications in the pathogenesis of hypertension. Angiotensin II-stimulated NAD(P)H oxidase in vascular cells results in the univalent reduction of O_2 in the presence of a free electron to yield superoxide anion ($\cdot O_2^-$), which in turn is dismutated to hydrogen peroxide (H_2O_2). In the presence of nitric oxide (NO), peroxynitrite ($ONOO^-$) can be formed. Increased intracellular levels of reactive oxygen species contribute to vascular inflammation, growth, altered contraction and dilation (vascular tone), and endothelial dysfunction. These events in turn lead to vascular remodeling, arterial narrowing, increased peripheral resistance and consequently to increased blood pressure. BH_4, tetrahydrobiopterin; eNOS, endothelial nitric oxide synthase; GPX, glutathione peroxidase; GSSG, oxidized glutathione; $\cdot OH^-$, hydroxyl radical; SOD, superoxide dismutase; TRXP, thioredoxin peroxidase; TRX-S_2, thioredoxin disulphide. Used with permission from *Can J Cardiol.* 2006; 22: 947–951 (Figure 1).

importantly, they exhibit an enhanced antioxidant capacity in adulthood [200]. Thus, these findings suggest that oxidative stress contributes to the etiology of hypertension programmed by *in utero* insult in male growth restricted offspring, yet protection against increased oxidative stress in female growth restricted offspring may contribute to the sex-specific fetal programming of hypertension.

5.8.3 Sex Differences in Oxidative Stress and the Role of the RAS

Experimental models of hypertension indicate that oxidative stress plays a more important role in mediating hypertension in male rats compared to the female [203]. Vascular superoxide production is greater in the male spontaneously hypertensive rat (SHR), a genetic model of experimental hypertension, relative to the female in a manner that is AT1R-dependent [204]. Angiotensin II-induced hypertension, an experimental model of hypertension, is more prevalent in the male rat or mouse relative to the female 204 [205]. The sex difference in response to angiotensin II is associated with a marked increase in renal superoxide production in the male and a marked increase in antioxidant capacity in the female [205]. Angiotensin II increases oxidative stress via AT1R-mediated stimulation of nicotinamide adenine dinucleotide phosphate (NAD(P)H) oxidase [206], indicating that the RAS is a source for increased oxidative stress. In an experimental model of fetal programming, hypertension-increased superoxide production observed in offspring exposed to global nutrient restriction during fetal life is NADPH oxidase-dependent [207]. Furthermore, the increase in oxidative stress in this model is abolished by blockade of the RAS [207], suggesting a key role for the RAS in the developmental programming of increased oxidative stress. Thus, these studies indicate that developmental programming of hypertension involves a key role of oxidative stress that may be angiotensin II-dependent.

5.9   GLUCOCORTICOIDS

The importance of fetal exposure to maternal glucocorticoids is demonstrated in studies whereby inhibition of placental 11β-hydroxysteroid dehydrogenase 2 (11-beta HSD2), the enzyme that protects the fetus from exposure to maternal glucocorticoids, programs growth restriction and hypertension in adult offspring [208]. Prenatal exposure to glucocorticoids administered to the pregnant rat during gestation also results in hypertension in adult rat offspring [76]. Yet, maternal glucocorticoid administration reduces food intake, leading to weight loss in the dam or pregnant rat [209]. Offspring of pair-fed dams (matching food intake in one group of pregnant rats to the food intake of the glucocorticoid treated pregnant rats) also demonstrate hypertension in later life [209], indicating that programming of hypertension via maternal glucocorticoids may involve the influence of maternal dietary intake and may not be the direct consequence of glucocorticoid exposure per se. A reduction in placental 11-beta HSD-2 expression and activity is reported in the placentas from low birth weight infants [210]. Additionally, a reduction in placental 11-beta HSD2 is observed in experimental models of IUGR induced by exposure to maternal low protein [211] and prenatal nicotine [212]. Blockade of fetal exposure to maternal glucocorticoids in the maternal protein restriction model of programmed hypertension prevents the development of hypertension in the offspring [213], indicating that programming by glucocorticoids contributes to the developmental programming of hypertension in this model. In addition, renal 11-beta HSD2 is reduced at birth in offspring of reduced uterine perfusion pregnant rats [214] indicating a role for fetal exposure to glucocorticoids as a potential mediator in the programming of later cardiovascular disease. Thus, these findings indicate that complications during pregnancy that result in slow fetal growth are associated with increased risk of fetal exposure to maternal glucocorticoids and implicate a role for glucocorticoids in the programming of later cardiovascular health.

5.10 EPIGENETIC MODIFICATIONS

Epigenetic processes modulate gene expression without altering the DNA sequence through mechanisms that involve changes in DNA methylation, histone modifications or through the influence of microRNAs [215] (Figure 14). Despite significant interest in the contribution of epigenetic mechanisms to the developmental programming of hypertension and chronic disease, few studies have directly tested the role of epigenetics in the programming of later chronic disease. Investigation, although limited, suggests that insults that occur during key periods of development result

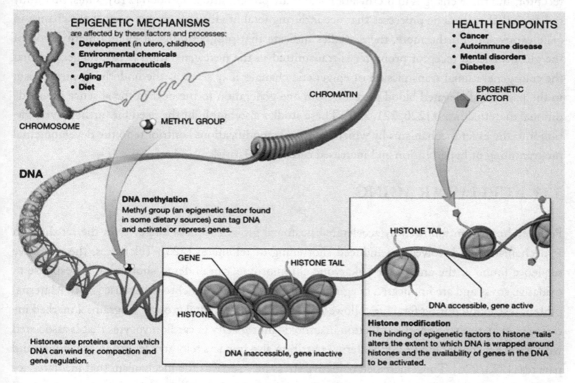

FIGURE 14: Schematic of the epigenetic mechanisms, DNA methylation and histone acetylation, associated with health and disease. Used with permission from the *NIH Roadmap (http://nihroadmap.nih .bov/EPIGENOMICS/epigeneticmecahanisms.asp).*

in epigenetic changes in genes that may be key to programming of later chronic disease. Maternal protein restriction in the rat results in changes in the methylation status of the adrenal AT1bR during fetal development [216]. Yet, whether these changes persist beyond fetal life and impact later chronic health is not yet known. A change in methylation of P53, a regulator of apoptosis genes in the kidney, is altered in the fetal response to placental insufficiency in the rat [217]. Apoptosis plays a key role in nephrogenesis. Thus, findings from this study suggest that epigenetic modifications that occur during fetal life may contribute to the reduction in nephron number observed in this model of developmental insult. Maternal protein restriction also alters the methylation pattern of the ACE gene in the fetal brain and programs up-regulation of miRNAs that regulate ACE mRNA translation [173]. A reduction in miRNAs that regulate AT2 translation are also observed in the fetal brain of offspring from protein-restricted dams [174] suggesting that fetal programming may involve epigenetic programming of factors key to cardiovascular health in later life. Offspring of a protein-restricted pregnant rat exhibit alterations in the methylation pattern of the glucocorticoid receptor, and these changes in methylation status are present into adulthood [218]. Thus, this study indicates that epigenetic processes that occur during fetal life may program a permanent change in gene expression. Furthermore, these studies indicate that programming of altered methylation of the glucocorticoid receptor promoter is transmitted to the next generation [219], suggesting that the transgenerational transmission of epigenetic changes may serve as the underlying mechanism in the passage of elevated blood pressure from one generation to the next in the absence of an additional maternal insult [220, 221, 222]. These studies also highlight the need for further investigation into the exact mechanisms by which epigenetic modifications contribute to the developmental programming of hypertension and increased cardiovascular risk.

5.11   CELLULAR AGING

Poor fetal nutrition followed by accelerated postnatal growth reduces longevity in the rat through a mechanism that involves age-induced shortening of telomeres [223]. Telomeres, the repetitive sequence found at the end of the eukaryotic chromosome, can undergo shortening in response to oxidative stress and are implicated in aging and the development of chronic disease [224]. Maternal protein restriction during fetal life followed by accelerated postnatal growth programs a marked increase in oxidative stress and a reduction in antioxidant capacity in cardiomyocytes that is associated with telomere shortening [225]. Maternal nutrition also impacts telomere length in the kidney and pancreas [223, 226]. Thus, increased oxidative stress may serve as the mechanism that mediates accelerated shortening of telomere length leading to later risk for cardiovascular disease programmed by in utero insult. Whether telomere length contributes to developmental programming of chronic disease in humans is not clear. Birth weight is not associated with shorter leucocyte telomere length in humans [227]. Yet, shorter leucocyte telomere length is observed in newborns and adults exposed to maternal stress during prenatal life [228, 229].

5.12 EARLY ACCELERATED GROWTH

Accelerated growth can follow undernutrition during fetal life if nutritional access is improved [230] and can be associated with short-term benefits as indicated by a reduction in illness and infant mortality during the first year of life [231]. However, accelerated growth in early postnatal life when preceded by low birth weight is associated with a significant risk of adult obesity, hypertension, cardiovascular and renal disease [12, 232]. Accelerated postnatal growth during the first 6 months of life serves as the critical window that predicts later adult blood pressure in infants born small [233]. Experimental studies also indicate that the timing of catch-up growth is key to the development of chronic disease in later life [234]. In addition, the impact of early postnatal accelerated catch-up growth following IUGR on later cardiovascular risk can be sex-specific [130]. Thus, critical windows of development are sensitive to adverse influences and span periods during fetal and early postnatal life. In addition, slow growth during fetal life followed by early postnatal accelerated growth exacerbates the impact of a poor fetal environment suggesting that a mismatch of growth during pre-and post-natal life represents a disruption in the capacity of the undernourished fetus to respond to increased metabolic demand after birth.

5.13 SUMMARY

The etiology of hypertension, a major risk factor for cardiovascular disease, is multifactorial. Insight from experimental studies of developmental programming highlights the complexity of the mechanisms involved in the increase in cardiovascular risk programmed in response to an insult during fetal and early postnatal life. The mechanistic pathways involved in the developmental programming of cardiovascular disease include alterations in the regulatory systems key for the long-term control of blood pressure. Similar alterations in these regulatory pathways are observed despite the different species or method used to induce a developmental insult, suggesting a key role for these regulatory systems in the programming of increased cardiovascular risk. Thus, insight from experimental studies suggests that hypertension programmed by a developmental insult may involve a reduction in GFR, an increase in tubular reabsorption, or both (Figure 15). Numerous regulatory systems contribute to the maintenance of sodium homeostasis and alterations in these systems such as the RAS and/or the SNS can impair the kidneys' ability to maintain body fluid and electrolyte homeostasis leading to hypertension. Importantly, sex and age can impact these regulatory systems and enhance or ameliorate the developmental programming of chronic disease.

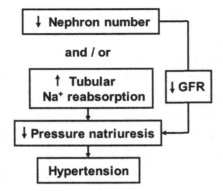

FIGURE 15: Potential renal mechanisms whereby developmental programming in response to *in utero* insult leads to development of hypertension. Used with permission from *Hypertension*. 2008 July; 52(1): 44–50 (Figure 2).

CHAPTER 6

Birth Weight and Clinical Considerations

6.1 BIRTH WEIGHT, WOMEN'S HEALTH, AND CHRONIC DISEASE

Cardiovascular disease is a leading cause of death in women in the Western world. In general, women develop cardiovascular disease later than men [235]. Whether changes in sex hormones precede or alter cardiovascular risk is not clear [for review see 235]. The attenuation of the age-related sex difference in cardiovascular risk may be better explained by the decrease in mortality that occurs in men at the age of normal female menopause rather than the impact of menopause on female health [236]. Few population studies are examining the impact of low birth weight on chronic health in postmenopausal women. Birth weight is inversely associated with blood pressure in women at 60 years of age regardless of BMI, but birth weight is not associated with blood pressure at 50 years of age, suggesting that age impacts the effect of slow fetal growth on blood pressure in women [59]. Low birth weight when coupled to adult obesity is associated with metabolic syndrome in post-menopausal women [237]. Thus, the risk for chronic disease in later life is increased in low birth weight women. Whether age or changes in estradiol or testosterone contribute is not clear.

Early onset menopause (<42 years of age) is positively associated with an increased risk for cardiovascular disease [238]. The mechanisms that mediate increased risk following early onset menopause are not known; yet, premature menopause is associated with higher BMI [239], a risk factor for cardiovascular disease. The causes of premature menopause are varied and include genetics, infection, autoimmune or metabolic disorders, and surgical interventions [240]. Early menopause may also be related to the age of onset in the mother [241] or ethnicity [242]. Low birth weight is associated with an increased risk for early menopause in the Sister Study, a prospective cohort study of US and Puerto Rican women aged 35–74 years [243]. Studies investigating the link between birth weight and age of menopause are lacking. Furthermore, the mechanisms linking low birth weight with early menopause are not clear but may be related to a reduction in primordial follicles [244]. Thus, these studies highlight the significant need to investigate the importance of age and menopausal transitions on chronic health in low birth weight women. In addition, these studies indicate that birth weight should be a consideration when implicating treatment options in the management of chronic disease.

Low birth weight women are at increased risk for complications during pregnancy including preeclampsia [68, 69], gestational diabetes [69, 70], and preterm delivery [71], suggesting that a woman's weight at birth impacts her gestational health in later life. Obesity increases the risk for a complicated pregnancy [245] and significantly exacerbates the risk for preeclampsia in women born low birth weight [246]. Experimental studies are limited but suggest that pregnancy in rats born growth restricted is associated with an increase in blood pressure that is impacted by advanced maternal age [247]. Pregnancy, which can serve as a natural stressor, also impacts cardiovascular [248] and metabolic risk [2479] in later life in female growth restricted rats relative to age-matched growth-restricted counterparts that do not experience pregnancy. Thus, low birth weight may serve as a risk factor for a complicated pregnancy and pregnancy may impact later chronic health in an individual born small.

In light of studies implicating preterm birth as a risk factor for later chronic health, early elective delivery, or birth scheduled without a medical reason prior to 39 weeks gestation via induction or Cesarean section, may have implications for future cardiovascular risk in the offspring. Early elective delivery is unnecessary according to the March of Dimes, American College of Obstetricians and Gynecologists, and the Women's Health Obstetric and Neonatal Nurses. In 2009, one in four births in the US was an early, elective delivery [250]. The rate of cesarean delivery can vary greatly [251]; however, changes in obstetric practice guidelines to avoid early elective delivery may greatly facilitate improved long-term health of future generations.

6.2   BIRTH WEIGHT, RACIAL DISPARITIES, AND CHRONIC DISEASE

Whites have a higher life expectancy than African Americans [252]. Racial disparities in cardiovascular and chronic renal disease implicate greater risk within the African American population [253, 254]. The exact mechanisms that contribute to racial disparities in chronic disease are not clear; however, risk factors for cardiovascular and chronic renal disease, such as BMI, are elevated in African Americans [255, 256]. The risk of low birth weight is two-fold higher in the African American population relative to the Caucasian population [257] (Figure 16). Thus, low birth weight, a significant indicator of increased cardiovascular and renal risk, may contribute to the racial disparities observed in chronic health. These findings highlight the importance of investigation into

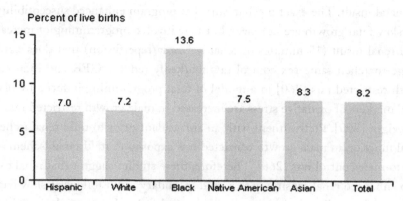

Low birthweight by race/ethnicity
United States, 2008-2010 Average

FIGURE 16: Low birth weight by race/ethnicity in the United States, 2008–2010 average. All race categories exclude Hispanics. Low birth weight is less than 5.5 pounds (2500 g). Used with permission from the March of Dimes.

the mechanisms linking fetal life with later chronic disease and implicate the importance of birth weight in the assessment of future health risk.

6.3 BIRTH WEIGHT AND THE MANAGEMENT OF CHRONIC DISEASE

Epidemiological and experimental evidence indicates that slow fetal growth enhances sensitivity to secondary influences. Accelerated early growth serves as a second insult after fetal growth restriction impacting the severity of long-term chronic health [12, 232]. Diabetic nephropathy and hypertension serve as major risk factors for end-stage renal disease [258]. However, the onset of end-stage renal disease is more rapid in hypertensive and/or diabetic low birth weight individuals relative to normal birth weight counterparts, suggesting that slow fetal growth programs enhanced susceptibility to renal injury in response to a secondary insult such as hypertension and diabetes [64, 65]. A reduction in nephron number is observed in low birth weight individuals [145], suggesting that a deficient in nephron number that occurs during fetal life may contribute to enhanced severity of renal disease in low birth weight individuals. However, a reduction in nephron number induced by gestational exposure to maternal protein restriction in an experimental model of developmental programming does not enhance renal dysfunction in response to hypoglycemia in the rat [259]. Thus, a reduction in nephron number per se, may not be sufficient to increase renal susceptibility to an additional post-natal insult. The exact mechanisms that program enhanced susceptibility to a second hit following slow fetal growth are not clear, but may involve programming of increased oxidative stress. A mild renal insult (15 minutes of renal ischemia/reperfusion) that does not impact renal function in age-matched same-sex control rats markedly reduces GFR and induces renal injury in male growth restricted rats [260] in a model of fetal programming induced by placental insufficiency. Renal markers of oxidative stress are increased in male growth restricted rats in this model of low birth weight [200]. Pretreatment with an antioxidant prior to mild renal ischemia prevents impaired renal function in male growth restricted rats exposed to mild renal ischemia normalizing GFR relative to male control rats [260]. Therefore, these studies suggest that renal oxidative stress contributes to enhanced renal sensitivity to a mild secondary insult following slow fetal growth.

Risk factors for cardiovascular disease are elevated in low birth weight individuals. Many of these risk factors including obesity, cholesterol and triglycerides, and blood pressure are modifiable in later life. However, in assessing a person's risk for cardiovascular disease, birth weight is not a modifiable risk factor. Clinical practice guidelines are available for the management of a pregnancy complicated by preeclampsia, gestational diabetes or IUGR, risk factors that impact fetal growth and later chronic health. Clinical practice guidelines are also available for the management of the preterm or the low birth weight (>2500 g) newborn. Yet, birth weight is not a consideration in the clinical practice guidelines for the management of hypertension, chronic kidney disease, stroke or heart disease in later life. Thus, birth weight, as an indicator of poor fetal growth, is linked to increased susceptibility to secondary influences that would not ordinarily impact chronic health. These findings suggest that birth weight may impact pharmacological and lifestyle interventions implicating the consideration of an individual's birth weight in the clinical management of cardiovascular disease.

6.4   SUMMARY

Numerous epidemiological studies implicate an inverse relationship between birth weight and blood pressure that is now widely accepted. Experimental models are providing insight into underlying mechanisms. However, based on the complexity of the etiology of hypertension and cardiovascular disease per se, the mechanisms that link early life with increased cardiovascular risk are multifactorial and confounded by increased sensitivity to secondary influences. Investigation into sex differences in the developmental programming of cardiovascular health in the human population is very limited. This is confounded by the lack of clarity related to the exact mechanisms that mediate sex and age-dependent differences in overall risk for hypertension, ischemic stroke, and coronary heart disease in the general population and the exact contribution of estrogen versus testosterone to increased cardiovascular disease following post-menopause [235]. The importance of birth weight as a clinical consideration in the management of a patient's heath is not currently considered in clinical practice guidelines for the management of chronic disease. However, the extensive number of studies that highlight the importance of early life influences on later chronic health suggest that birth weight, as a marker of early life events, should be a consideration in the assessment of treatment and interventional strategies in health care and lifestyle choices.

·   ·   ·   ·   ·

CHAPTER 7

Birth Weight and Cardiovascular Disease: Translational Considerations

- Should birth weight be a consideration in the interventional or therapeutic approach in chronic disease management?
- How does enhanced sensitivity to a secondary influence that is not relevant in the general population alter disease outcome and management of chronic disease in a low birth weight individual?
- Does age impact events programmed in early life and augment the progression and severity of chronic disease?
- What impact does birth weight have on chronic health in postmenopausal women?
- What impact does birth weight have on gestational health in women born preterm or small for gestational age?
- What impact does low birth weight have on racial disparities in chronic health?

. . . .

CHAPTER 8

Summary and Conclusions

8.1 OVERVIEW OF UNDERLYING MECHANISMS

Experimental models of developmental programming of cardiovascular disease are providing insight into the underlying mechanisms by which slow fetal growth programs high blood pressure and increased cardiovascular risk. The causative factors are multifactorial and the exact interactions that mediate a reduction in nephron number, hypertension and increased cardiovascular and renal risk in the offspring are not yet clearly elucidated.

8.2 DEVELOPMENTAL PROGRAMMING AND NEPHRON NUMBER

Undernutrition in fetal life programs a marked decrease in renal renin, angiotensinogen [75, 162], and COX-2 expression [214], critical mediators of proper nephrogenesis. A decrease in the expression of renal renin, angiotensinogen, and COX-2 are noted in numerous experimental models of developmental insult that exhibit a reduction in nephron number [75, 160, 162]. Glucocorticoids contribute to the regulation of the RAS and COX-2 during development [216] and reductions in gene expression may involve changes in DNA methylation linked to fetal exposure to maternal glucocorticoids [259]. Thus, suppression of the renal RAS and COX-2 during nephrogenesis may be an underlying mechanism in the programming of impaired nephrogenesis and reduced nephron number. Moreover, fetal exposure to glucocorticoids may induce epigenetic processes that impact expression of these key factors [174, 261] and contribute to altered organogenesis and the programming of chronic disease such as hypertension in later life (Figure 17).

8.3 DEVELOPMENTAL PROGRAMMING AND HYPERTENSION

Undernutrition increases expression of the RAS in regions of the brain key to cardiovascular control [173, 174]. Central actions of the RAS can include modulation of the SNS, with the renal nerves

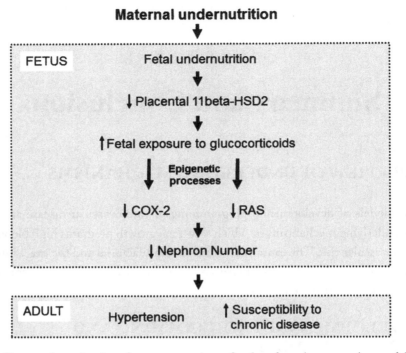

FIGURE 17: Proposed mechanism for programming of reduced nephron number and hypertension. Undernutrition during fetal life leads to excess exposure of the fetus to maternal glucocorticoids, altering renal expression of renin and COC-2 via epigenetic mechanisms. Reductions in renin and COX-2, key mediators of nephrogenesis, cause a reduction in nephron number leading to increased risk for chronic disease in later life.

serving as the link between the SNS and the kidney [178]. Several studies implicate the renal nerves in the etiology of hypertension induced by prenatal insult [137, 181, 182]. Thus, inappropriate activation of the central RAS [175], leading to an increase in renal sympathetic nerve activity, may be one underlying mechanism by which insults during fetal life program hypertension.

Blockade of the systemic RAS also abolishes hypertension induced by *in utero* exposure to undernutrition [153, 154, 158]. The RAS is a key mediator of the long-term control of blood pressure, and inappropriate activation of the RAS is observed in many experimental models of hypertension [262]. Whether the contribution of the RAS to programmed hypertension originates from circulating or intrarenal sources of the RAS is not yet clear. However, all components of the RAS are expressed within the kidney [263] and the importance of intrarenal RAS in the absence of changes in circulating RAS as a mediator of hypertension is well established [264]. ANG II can increase oxi-

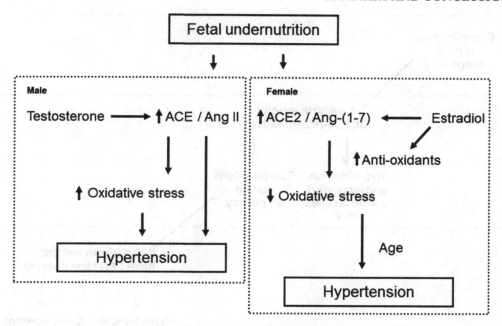

FIGURE 18: Overview of potential underlying mechanisms in the developmental programming of hypertension by undernutrition during fetal life.

dative stress [265]. Thus, the mechanisms by which oxidative stress is increased in models induced by placental insufficiency and maternal undernutrition [200, 201, 202] could involve the RAS. Sex hormones differentially modulate expression of vasoconstrictor or vasodilator pathways of the RAS [166, 167, 170, 171], and a key role for the sex hormones is noted in many experimental models of developmental insult [118, 142, 154]. Age also impacts programmed risk [137], suggesting that the origins of developmental programming of hypertension and increased cardiovascular risk may be multifactorial (Figure 18). These developmental origins may involve numerous interactions between different regulatory pathways key to the long-term control of blood pressure and additional factors, such as age and sex hormones, may greatly influence later chronic health and disease.

8.4   DEVELOPMENTAL PROGRAMMING AND INTERGENERATIONAL EFFECTS

Adverse influences during early development such as exposure to maternal protein restriction during pregnancy or uteroplacental insufficiency increases blood pressure and induces vascular dysfunction in the F1 and F2 generation [194, 222]. Importantly, transmission of programmed cardiovascular

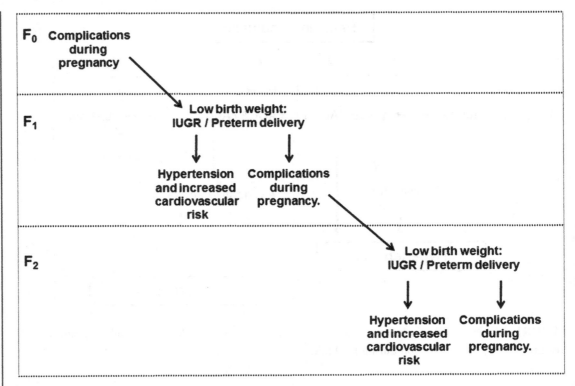

FIGURE 19: Schematic highlighting how influences during fetal life program increased gestational and cardiovascular risk from one generation to the next. Intrauterine growth restriction (IUGR).

risk occurs in the absence of an additional maternal insult. Fetal growth restriction impacts uterine and placental size [266], implicating a potential mechanism by which programmed risk may be propagated to a subsequent generation. Low birth weight is also a risk factor for complications during pregnancy such as preeclampsia which can induce low birth weight of the offspring [68, 69]. Epigenetic processes may also contribute to the transgenerational passage of increased cardiovascular risk [219]. Thus, the impact of growth during early life in one generation can greatly impact the gestational health of the next generation and the chronic health of subsequent generations (Figure 19). These studies implicate the importance of fetal growth and organogenesis in one generation on the long-term chronic health of the next.

References

[1]   Foraker RE, Rose KM, Kucharska-Newton AM, Ni H, Suchindran CM, Whitsel EA. Variation in rates of fatal coronary heart disease by neighborhood socioeconomic status: the atherosclerosis risk in communities surveillance (1992–2002). *Ann Epidemiol.* 2011; 21: pp. 580–8.

[2]   Kucharska-Newton AM, Harald K, Rosamond WD, Rose KM, Rea TD, Salomaa V. Socioeconomic indicators and the risk of acute coronary heart disease events: comparison of population-based data from the United States and Finland. *Ann Epidemiol.* 2011; 21: pp. 572–9.

[3]   Rose G, Marmot MG. Social class and coronary heart disease. *Br Heart J.* 1981; 45: pp. 13–9.

[4]   Forsdahl A. Observations throwing light on the high mortality in the county of Finmark. Is the high mortality today a late effect of very poor living conditions in childhood and adolescence? 1973. *Int J Epidemiol.* 2002; 31: pp. 302–8.

[5]   Wadsworth ME, Cripps HA, Midwinter RE, Colley JR. Blood pressure in a national birth cohort at the age of 36 related to social and familial factors, smoking, and body mass. *Br Med J (Clin Res Ed).* 1985; 291: pp. 1534–8.

[6]   Barker DJ, Osmond C. Infant mortality, childhood nutrition, and ischaemic heart disease in England and Wales. *Lancet.* 1986; 1: pp. 1077–81.

[7]   Barker DJ, Osmond C. Low birth weight and hypertension. *BMJ.* 1988; 297: pp. 134–5.

[8]   Barker DJ, Osmond C, Golding J, Kuh D, Wadsworth ME. Growth in utero, blood pressure in childhood and adult life, and mortality from cardiovascular disease. *BMJ.* 1989; 298: pp. 564–7.

[9]   Hendrix N, Berghella V. Non-placental causes of intrauterine growth restriction. *Semin Perinatol.* 2008; 32: pp. 161–5.

[10]   Barker D. *Mother, Babies, and Disease in Later Life.* BMJ Publishing Group, London, 1994.

[11]   Barker DJ, Bagby SP. Developmental antecedents of cardiovascular disease: a historical perspective. *J Am Soc Nephrol.* 2005; 16: pp. 2537–44.

[12]   Singhal A, Cole TJ, Fewtrell M, Kennedy K, Stephenson T, Elias-Jones A, Lucas A. Promotion of faster weight gain in infants born small for gestational age: is there an adverse effect on later blood pressure? *Circulation.* 2007; 115: pp. 213–20.

[13]   Ong KK, Ahmed ML, Emmett PM, Preece MA, Dunger DB. Association between postnatal catch-up growth and obesity in childhood: prospective cohort study. *BMJ.* 2000; 320: pp. 967–71.

[14]   Jimenez-Chillaron JC, Patti ME. To catch up or not to catch up: is this the question? Lessons from animal models. *Curr Opin Endocrinol Diabetes Obes.* 2007; 14: pp. 23–9.

[15]   Bansal N, Ayoola OO, Gemmell I, Vyas A, Koudsi A, Oldroyd J, Clayton PE, Cruickshank JK. Effects of early growth on blood pressure of infants of British European and South Asian origin at one year of age: the Manchester children's growth and vascular health study. *J Hypertens.* 2008; 26: pp. 412–8.

[16]   Chang J, Streitman D. Physiologic adaptations to pregnancy. *Neurol Clin.* 2012; 30: pp. 781–9.

[17]   Romo A, Carceller R, Tobajas J. Intrauterine growth retardation (IUGR): epidemiology and etiology. *Pediatr Endocrinol Rev.* 2009; 6: pp. 332–6.

[18]   Davis EF, Lazdam M, Lewandowski AJ, Worton SA, Kelly B, Kenworthy Y, Adwani S, Wilkinson AR, McCormick K, Sargent I, Redman C, Leeson P. Cardiovascular risk factors in children and young adults born to preeclamptic pregnancies: a systematic review. *Pediatrics.* 2012; 129: pp. e1552–61.

[19]   Lawlor DA, Macdonald-Wallis C, Fraser A, Nelson SM, Hingorani A, Davey Smith G, Sattar N, Deanfield J. Cardiovascular biomarkers and vascular function during childhood in the offspring of mothers with hypertensive disorders of pregnancy: findings from the Avon Longitudinal Study of Parents and Children. *Eur Heart J.* 2012; 33: pp. 335–45.

[20]   Lazdam M, de la Horra A, Pitcher A, Mannie Z, Diesch J, Trevitt C, Kylintireas I, Contractor H, Singhal A, Lucas A, Neubauer S, Kharbanda R, Alp N, Kelly B, Leeson P. Elevated blood pressure in offspring born premature to hypertensive pregnancy: is endothelial dysfunction the underlying vascular mechanism? *Hypertension.* 2010; 56: pp. 159–65.

[21]   West NA, Crume TL, Maligie MA, Dabelea D. Cardiovascular risk factors in children exposed to maternal diabetes in utero. *Diabetologia.* 2011; 54: pp. 504–7.

[22]   Mamun AA, O'Callaghan MJ, Williams GM, Najman JM. Maternal smoking during pregnancy predicts adult offspring cardiovascular risk factors—evidence from a community-based large birth cohort study. *PLoS One.* 2012; 7:e41106. doi: 10.1371/journal.pone.0041106.

[23]   Chen X, Wang Y. Tracking of blood pressure from childhood to adulthood: a systematic review and meta-regression analysis. *Circulation.* 2008; 117: pp. 3171–80.

[24] Primatesta P, Falaschetti E, Poulter NR. Birth weight and blood pressure in childhood: results from the Health Survey for England. *Hypertension.* 2005; 45: pp. 75–9.

[25] Strufaldi MW, Silva EM, Franco MC, Puccini RF. Blood pressure levels in childhood: probing the relative importance of birth weight and current size. *Eur J Pediatr.* 2009; 168: pp. 619–24.

[26] Johnson RC, Schoeni RF. Early-life origins of adult disease: national longitudinal population-based study of the United States. *Am J Public Health.* 2011; 101: pp. 2317–24.

[27] Spence D, Stewart MC, Alderdice FA, Patterson CC, Halliday HL. Intra-uterine growth restriction and increased risk of hypertension in adult life: a follow-up study of 50-year-olds. *Public Health.* 2012; 126: pp. 561–5.

[28] Law CM, de Swiet M, Osmond C, Fayers PM, Barker DJ, Cruddas AM, Fall CH. Initiation of hypertension in utero and its amplification throughout life. *BMJ.* 1993; 306: pp. 24–7.

[29] Eriksson JG, Forsén T, Tuomilehto J, Osmond C, Barker DJ. Early growth and coronary heart disease in later life: longitudinal study. *BMJ.* 2001; 322: pp. 949–53.

[30] Grove JS, Reed DM, Yano K, Hwang LJ. Variability in systolic blood pressure—a risk factor for coronary heart disease? *Am J Epidemiol.* 1997; 145: pp. 771–6.

[31] Chen W, Srinivasan SR, Yao L, Li S, Dasmahapatra P, Fernandez C, Xu J, Berenson GS. Low birth weight is associated with higher blood pressure variability from childhood to young adulthood: the Bogalusa Heart Study. *Am J Epidemiol.* 2012; 176: pp. S99–105.

[32] Sowers JR. Obesity as a cardiovascular risk factor. *Am J Med.* 2003; 115: pp. 37S–41S.

[33] Stein AD, Kahn HS, Rundle A, Zybert PA, van der Pal-de Bruin K, Lumey LH. Anthropometric measures in middle age after exposure to famine during gestation: evidence from the Dutch famine. *Am J Clin Nutr.* 2007; 85: pp. 869–76.

[34] Lumey LH. Decreased birth weights in infants after maternal in utero exposure to the Dutch famine of 1944–1945. *Paediatr Perinat Epidemiol.* 1992; 6: pp. 240–53.

[35] Skilton MR, Viikari JS, Juonala M, Laitinen T, Lehtimäki T, Taittonen L, Kähönen M, Celermajer DS, Raitakari OT. Fetal growth and preterm birth influence cardiovascular risk factors and arterial health in young adults: the Cardiovascular Risk in Young Finns Study. *Arterioscler Thromb Vasc Biol.* 2011; 31: pp. 2975–81.

[36] Davies AA, Smith GD, Ben-Shlomo Y, Litchfield P. Low birth weight is associated with higher adult total cholesterol concentration in men: findings from an occupational cohort of 25,843 employees. *Circulation.* 2004; 110: pp. 1258–62.

[37] Kvehaugen AS, Dechend R, Ramstad HB, Troisi R, Fugelseth D, Staff AC. Endothelial function and circulating biomarkers are disturbed in women and children after preeclampsia. *Hypertension.* 2011; 58: pp. 63–9.

[38]   Leeson CP, Kattenhorn M, Morley R, Lucas A, Deanfield JE. Impact of low birth weight and cardiovascular risk factors on endothelial function in early adult life. *Circulation*. 2001; 103: pp. 1264–8.

[39]   Berends LM, Ozanne SE. Early determinants of type-2 diabetes. *Best Pract Res Clin Endocrinol Metab*. 2012; 26: pp. 569–80.

[40]   White SL, Perkovic V, Cass A, Chang CL, Poulter NR, Spector T, Haysom L, Craig JC, Salmi IA, Chadban SJ, Huxley RR. Is low birth weight an antecedent of CKD in later life? A systematic review of observational studies. *Am J Kidney Dis*. 2009; 54: pp. 248–61.

[41]   de Jong F, Monuteaux MC, van Elburg RM, Gillman MW, Belfort MB. Systematic review and meta-analysis of preterm birth and later systolic blood pressure. *Hypertension*. 2012; 59: pp. 226–34.

[42]   Singhal A, Cole TJ, Lucas A. Early nutrition in preterm infants and later blood pressure: two cohorts after randomised trials. *Lancet*. 2001; 357: pp. 413–9.

[43]   Kerkhof GF, Willemsen RH, Leunissen RW, Breukhoven PE, Hokken-Koelega AC. Health profile of young adults born preterm: negative effects of rapid weight gain in early life. *J Clin Endocrinol Metab*. 2012; 97: pp. 4498–506.

[44]   Johansson S, Iliadou A, Bergvall N, Tuvemo T, Norman M, Cnattingius S. Risk of high blood pressure among young men increases with the degree of immaturity at birth. *Circulation*. 2005 29; 112: pp. 3430–6.

[45]   Breukhoven PE, Kerkhof GF, Willemsen RH, Hokken-Koelega AC. Fat mass and lipid profile in young adults born preterm. *J Clin Endocrinol Metab*. 2012; 97: pp. 1294–302.

[46]   Kerkhof GF, Breukhoven PE, Leunissen RW, Willemsen RH, Hokken-Koelega AC. Does preterm birth influence cardiovascular risk in early adulthood? *J Pediatr*. 2012; 161: pp. 390–6.e1.

[47]   Finken MJ, Keijzer-Veen MG, Dekker FW, Frölich M, Hille ET, Romijn JA, Wit JM, Dutch POPS-19 Collaborative Study Group. Preterm birth and later insulin resistance: effects of birth weight and postnatal growth in a population based longitudinal study from birth into adult life. *Diabetologia*. 2006; 49: pp. 478–5.

[48]   Dalziel SR, Parag V, Rodgers A, Harding JE. Cardiovascular risk factors at age 30 following pre-term birth. *Int J Epidemiol*. 2007; 36: pp. 907–15.

[49]   Weisfeldt ML, Zieman SJ. Advances in the prevention and treatment of cardiovascular disease. *Health Affairs*. 2007; 26: pp. 25–37.

[50]   Khoury S, Yarows SA, O'Brien TK, Sowers JR. Ambulatory blood pressure monitoring in a nonacademic setting. Effects of age and sex. *Am J Hypertens*. 1992; 5: pp. 616–23.

[51]   Winberg N, Hoegholm A, Christensen HR, Bang LE, Mikkelsen KL, Nielsen PE, Svendsen TL, Kampmann JP, Madsen NH, Bentzon MW. 24-h ambulatory blood pressure in 352 normal Danish subjects, related to age and gender. *Am J Hypertens*. 1995; 8: pp. 978–86.

[52] Law CM, Shiell AW, Newsome CA, Syddall HE, Shinebourne EA, Fayers PM, Martyn CN, de Swiet M. Fetal, infant, and childhood growth and adult blood pressure: a longitudinal study from birth to 22 years of age. *Circulation*. 2002; 105: pp. 1088–92.

[53] Moore VM, Cockington RA, Ryan P, Robinson JS. The relationship between birth weight and blood pressure amplifies from childhood to adulthood. *J Hypertens*. 1999; 17: pp. 883–8.

[54] Gamborg M, Byberg L, Rasmussen F, Andersen PK, Baker JL, Bengtsson C, Canoy D, Drøyvold W, Eriksson JG, Forsén T, Gunnarsdottir I, Järvelin MR, Koupil I, Lapidus L, Nilsen TI, Olsen SF, Schack-Nielsen L, Thorsdottir I, Tuomainen TP, Sørensen TI, Nord-Net Study Group. Birth weight and systolic blood pressure in adolescence and adulthood: meta-regression analysis of sex- and age-specific results from 20 Nordic studies. *Am J Epidemiol*. 2007; 166: pp. 634–45.

[55] Jones A, Beda A, Osmond C, Godfrey KM, Simpson DM, Phillips DI. Sex-specific programming of cardiovascular physiology in children. *Eur Heart J*. 2008; 29: pp. 2164–70.

[56] Vos LE, Oren A, Bots ML, Gorissen WH, Grobbee DE, Uiterwaal CS. Birth size and coronary heart disease risk score in young adulthood. The Atherosclerosis Risk in Young Adults (ARYA) study. *Eur J Epidemiol*. 2006; 21: pp. 33–8.

[57] Hallan S, Euser AM, Irgens LM, Finken MJ, Holmen J, Dekker FW. Effect of intrauterine growth restriction on kidney function at young adult age: the Nord Trondelag Health (HUNT 2) Study. *Am J Kidney Dis*. 2008; 51: pp. 10–20.

[58] Jarvelin MR, Sovio U, King V, Lauren L, Xu B, McCarthy MI, Hartikainen AL, Laitinen J, Zitting P, Rantakallio P, Elliott P. Early life factors and blood pressure at age 31 years in the 1966 northern Finland birth cohort. *Hypertension*. 2004; 44: pp. 838–46.

[59] Andersson SW, Lapidus L, Niklasson A, Hallberg L, Bengtsson C, Hulthén L. Blood pressure and hypertension in middle-aged women in relation to weight and length at birth: a follow-up study. *J Hypertens*. 2000; 18: pp. 1753–61.

[60] Parkinson JR, Hyde MJ, Gale C, Santhakumaran S, Modi N. Preterm birth and the metabolic syndrome in adult life: a systematic review and meta-analysis. *Pediatrics*. 2013; 13: pp. e1240–63.

[61] Kistner A, Jacobson L, Jacobson SH, Svensson E, Hellstrom A. Low gestational age associated with abnormal retinal vascularization and increased blood pressure in adult women. *Pediatr Res*. 2002; 51: pp. 675–80.

[62] Bonamy AK, Bendito A, Martin H, Andolf E, Sedin G, Norman M. Preterm birth contributes to increased vascular resistance and higher blood pressure in adolescent girls. *Pediatr Res*. 2005; 58: pp. 845–89.

[63] Kistner A, Celsi G, Vanpee M, Jacobson SH. Increased blood pressure but normal renal function in adult women born preterm. *Pediatr Nephrol*. 2000; 15: pp. 215–20.

[64] Lackland DT, Bendall HE, Osmond C, Egan BM, Barker DJ. Low birth weights

contribute to high rates of early-onset chronic renal failure in the Southeastern United States. *Arch Intern Med*. 2000; 160: pp. 1472–6.

[65]   Orskov B, Christensen KB, Feldt-Rasmussen B, Strandgaard S. Low birth weight is associated with earlier onset of end-stage renal disease in Danish patients with autosomal dominant polycystic kidney disease. *Kidney Int*. 2012; 81: pp. 919–24.

[66]   Zidar N, Avgustin Cavic M, Kenda RB, Ferluga D. Unfavorable course of minimal change nephrotic syndrome in children with intrauterine growth retardation. *Kidney Int*. 1998; 54: pp. 1320–3.

[67]   Bilhartz TD, Bilhartz PA, Bilhartz TN, Bilhartz RD. Making use of a natural stress test: pregnancy and cardiovascular risk. *J Womens Health (Larchmt)*. 2011; 20: pp. 695–701.

[68]   Innes KE, Marshall JA, Byers TE, Calonge N. A woman's own birth weight and gestational age predict her later risk of developing preeclampsia, a precursor of chronic disease. *Epidemiology*. 1999; 10: pp. 153–60.

[69]   Boivin A, Luo ZC, Audibert F, Masse B, Lefebvre F, Tessier R et al. Pregnancy complications among women born preterm. *CMAJ*. 2012; 184: pp. 1777–84.

[70]   Bo S, Marchisio B, Volpiano M, Menato G, Pagano G. Maternal low birth weight and gestational hyperglycemia. *Gynecol Endocrinol*. 2003; 7: pp. 133–6.

[71]   Innes KE, Byers TE, Marshall JA, Baron A, Orleans M, Hamman RF. Association of a woman's own birth weight with subsequent risk for gestational diabetes. *JAMA*. 2002; 287: pp. 2534–41.

[72]   De B, Lin S, Lohsoonthorn V, Williams MA. Risk of preterm delivery in relation to maternal low birth weight. *Acta Obstet Gynecol Scand*. 2007; 86: pp. 565–71.

[73]   Romundstad PR, Magnussen EB, Smith GD, Vatten LJ. Hypertension in pregnancy and later cardiovascular risk: common antecedents? *Circulation*. 2010; 122: pp. 579–84.

[74]   Langley-Evans SC, Phillips GJ, Jackson AA. In utero exposure to maternal low protein diets induces hypertension in weanling rats, independently of maternal blood pressure changes. *Clin Nutr*. 1994; 13: pp. 319–24.

[75]   Woods LL, Ingelfinger JR, Nyengaard JR, Rasch R. Maternal protein restriction suppresses the newborn renin-angiotensin system and programs adult hypertension in rats. *Pediatr Res*. 2001; 49: pp. 460–7.

[76]   Ortiz LA, Quan A, Zarzar F, Weinberg A, Baum M. Prenatal dexamethasone programs hypertension and renal injury in the rat. *Hypertension*. 2003; 41: pp. 328–34.

[77]   Williams SJ, Hemmings DG, Mitchell JM, McMillen IC, Davidge ST. Effects of maternal hypoxia or nutrient restriction during pregnancy on endothelial function in adult male rat offspring. *J Physiol*. 2005; 565: pp. 125–35.

[78]   Liu J, Gao Y, Negash S, Longo LD, Raj JU. Long-term effects of prenatal hypoxia on

endothelium-dependent relaxation responses in pulmonary arteries of adult sheep. *Am J Physiol Lung Cell Mol Physiol.* 2009; 296: pp. L547–54.

[79] Alexander, BT. Placental insufficiency leads to development of hypertension in growth restricted offspring. *Hypertension.* 2003; 41: pp. 457–62.

[80] Gomez RA, Norwood VF. Recent advances in renal development. *Curr Opin Pediatr.* 1999; 11: pp. 135–40.

[81] Zoetis T, Hurtt ME. Species comparison of anatomical and functional renal development. *Birth Defects Res B Dev Reprod Toxicol.* 2003; 68: pp. 111–20.

[82] Kawamura M, Itoh H, Yura S, Mogami H, Suga S, Makino H, Miyamoto Y, Yoshimasa Y, Sagawa N, Fujii S. Undernutrition in utero augments systolic blood pressure and cardiac remodeling in adult mouse offspring: possible involvement of local cardiac angiotensin system in developmental origins of cardiovascular disease. *Endocrinology.* 2007; 148: pp. 1218–25.

[83] Kind KL, Simonetta G, Clifton PM, Robinson JS, Owens JA. Effect of maternal feed restriction on blood pressure in the adult guinea pig. *Exp Physiol.* 2002; 87: pp. 469–77.

[84] Gilbert JS, Lang AL, Grant AR, Nijland MJ. Maternal nutrient restriction in sheep: hypertension and decreased nephron number in offspring at 9 months of age. *J Physiol* 2005; 565: pp. 137–47.

[85] Wlodek ME, Mibus A, Tan A, Siebel AL, Owens JA, Moritz KM. Normal lactational environment restores nephron endowment and prevents hypertension after placental restriction in the rat. *J Am Soc Nephrol.* 2007; 18: pp. 1688–96.

[86] Zohdi V, Moritz KM, Bubb KJ, Cock ML, Wreford N, Harding R, Black MJ. Nephrogenesis and the renal renin-angiotensin system in fetal sheep: effects of intrauterine growth restriction during late gestation. *Am J Physiol Regul Integr Comp Physiol.* 2007; 293: pp. R1267–73.

[87] Dickinson H, Walker DW, Wintour EM, Moritz K. Maternal dexamethasone treatment at midgestation reduces nephron number and alters renal gene expression in the fetal spiny mouse. *Am J Physiol Regul Integr Comp Physiol.* 2007; 292: pp. R453–61.

[88] Ortiz LA, Quan A, Weinberg A, Baum M. Effect of prenatal dexamethasone on rat renal development. *Kidney Int.* 2001; 59: pp. 1663–9.

[89] Wintour EM, Moritz KM, Johnson K, Ricardo S, Samuel CS, Dodic M. Reduced nephron number in adult sheep, hypertensive as a result of prenatal glucocorticoid treatment. *J Physiol.* 2003; 549: pp. 929–35.

[90] Roghair RD, Segar JL, Kilpatrick RA, Segar EM, Scholz TD, Lamb FS. Murine aortic reactivity is programmed equally by maternal low protein diet or late gestation dexamethasone. *J Matern Fetal Neonatal Med.* 2007; 20: pp. 833–41.

[91]   Holemans K, Gerber R, Meurrens K, De Clerck F, Poston L, Van Assche FA. Maternal food restriction in the second half of pregnancy affects vascular function but not blood pressure of rat female offspring. *Br J Nutr.* 1999; 81: pp. 73–7.

[92]   Torrens C, Snelling TH, Chau R, Shanmuganathan M, Cleal JK, Poore KR, Noakes DE, Poston L, Hanson MA, Green LR. Effects of pre- and periconceptional undernutrition on arterial function in adult female sheep are vascular bed dependent. *Exp Physiol.* 2009; 94: pp. 1024–33.

[93]   Guyton AC, Coleman TG, Cowley AV, Jr, et al. Arterial pressure regulation. Overriding dominance of the kidneys in long-term regulation and in hypertension. *Am J Med.* 1972; 52: pp. 584–94.

[94]   Woods LL, Rasch R. Perinatal ANG II programs adult blood pressure, glomerular number, and renal function in rats. *Am J Physiol.* 1998; 275: pp. R1593–9.

[95]   Loria A, Reverte V, Salazar F, Saez F, Llinas MT, Salazar FJ. Sex and age differences of renal function in rats with reduced ANG II activity during the nephrogenic period. *Am J Physiol Renal Physiol.* 2007; 293: pp. F506–10.

[96]   Bernstein I, Gabbe GS. Intrauterine growth restriction. Obstetrics, Normal and Problem Pregnancies. In: Gabbe SG, Neibyl JR, Simpson JL, Annas GJ editors. Churchill Livingstone, Philadelphia, PA: 2002. pp. 869–91.

[97]   Vehaskari VM, Aviles DH, Manning J. Prenatal programming of adult hypertension in the rat. *Kidney Int.* 2001; 59: pp. 238–45.

[98]   Cheema KK, Dent MR, Saini HK, Aroutiounova N, Tappia PS. Prenatal exposure to maternal undernutrition induces adult cardiac dysfunction. *Br J Nutr.* 2005; 93: pp. 471–7.

[99]   Ozaki T, Nishina H, Hanson MA, Poston L. Dietary restriction in pregnant rats causes gender-related hypertension and vascular dysfunction in offspring. *J Physiol.* 2001; 530: pp. 141–52.

[100] Gopalakrishnan GS, Gardner DS, Rhind SM, Rae MT, Kyle CE, Brooks AN, Walker RM, Ramsay MM, Keisler DH, Stephenson T, Symonds ME. Programming of adult cardiovascular function after early maternal undernutrition in sheep. *Am J Physiol Regul Integr Comp Physiol.* 2004; 287: pp. R12–20.

[101] Wigglesworth JS. Fetal growth retardation. Animal model: uterine vessel ligation in the pregnant rat. *J Path Biol.* 1964, 88, pp. 1–13.

[102] Louey S, Jonker SS, Giraud GD, Thornburg KL. Placental insufficiency decreases cell cycle activity and terminal maturation in fetal sheep cardiomyocytes. *J Physiol.* 2007; 580: pp. 639–48.

[103] Dodson RB, Rozance PJ, Fleenor BS, Petrash CC, Shoemaker LG, Hunter KS, Ferguson VL. Increased arterial stiffness and extracellular matrix reorganization in intrauterine growth-restricted fetal sheep. *Pediatr Res.* 2013; 73: pp. 147–54.

[104] George EM, Granger JP. Mechanisms and potential therapies for preeclampsia. *Curr Hypertens Rep*. 2011; 13: pp. 269–75.

[105] Xu Y, Williams SJ, O'Brien D, Davidge ST. Hypoxia or nutrient restriction during pregnancy in rats leads to progressive cardiac remodeling and impairs postischemic recovery in adult male offspring. *FASEB J*. 2006; 20: pp. 1251–3.

[106] Giussani DA, Camm EJ, Niu Y, Richter HG, Blanco CE, Gottschalk R, Blake EZ, Horder KA, Thakor AS, Hansell JA, Kane AD, Wooding FB, Cross CM, Herrera EA. Developmental programming of cardiovascular dysfunction by prenatal hypoxia and oxidative stress. *PLoS One*. 2012; 7:e31017. doi: 10.1371/journal.pone.0031017.

[107] Yzydorczyk C, Comte B, Cambonie G, Lavoie JC, Germain N, Ting Shun Y, Wolff J, Deschepper C, Touyz RM, Lelièvre-Pegorier M, Nuyt AM. Neonatal oxygen exposure in rats leads to cardiovascular and renal alterations in adulthood. *Hypertension*. 2008; 52: pp. 889–95.

[108] Chrousos GP, Kino T. Glucocorticoid signaling in the cell. Expanding clinical implications to complex human behavioral and somatic disorders. *Ann N Y Acad Sci*. 2009; 1179: pp. 153–66.

[109] Khulan B, Drake AJ. Glucocorticoids as mediators of developmental programming effects. *Best Pract Res Clin Endocrinol Metab*. 2012; 26: pp. 689–700.

[110] Herrera M, Coffman TM. The kidney and hypertension: novel insights from transgenic models. *Curr Opin Nephrol Hypertens*. 2012; 21: pp. 171–8.

[111] Guron G, Friberg P. An intact renin–angiotensin system is a prerequisite for normal renal development. *J Hypertens*. 2000; 18: pp. 123–37.

[112] Norwood VF, Morham SG, Smithies O. Postnatal development and progression of renal dysplasia in cyclooxygenase-2 null mice. *Kidney Int*. 2000; 58: pp. 2291–300.

[113] Sáez F, Reverte V, Salazar F, Castells MT, Llinás MT, Salazar FJ. Hypertension and sex differences in the age-related renal changes when cyclooxygenase-2 activity is reduced during nephrogenesis. *Hypertension*. 2009; 53: pp. 331–7.

[114] Blake KV, Gurrin LC, Evans SF, Beilin LJ, Landau LI, Stanley FJ, Newnham JP. Maternal cigarette smoking during pregnancy, low birth weight and subsequent blood pressure in early childhood. *Early Hum Dev*. 2000; 57: pp. 137–47.

[115] Xiao D, Xu Z, Huang X, Longo LD, Yang S, Zhang L. Prenatal gender-related nicotine exposure increases blood pressure response to angiotensin II in adult offspring. *Hypertension*. 2008; 51: pp. 1239–47.

[116] Dong M, Giles WH, Felitti VJ, Dube SR, Williams JE, Chapman DP, Anda RF. Insights into causal pathways for ischemic heart disease: adverse childhood experiences study. *Circulation*. 2004; 110: pp. 1761–6.

[117] Loria AS, Pollock DM, Pollock JS. Early life stress sensitizes rats to angiotensin

II-induced hypertension and vascular inflammation in adult life. *Hypertension*. 2010; 55: pp. 494–9.

[118] Loria AS, Yamamoto T, Pollock DM, Pollock JS. Early life stress induces renal dysfunction in adult male rats but not female rats. *Am J Physiol Regul Integr Comp Physiol*. 2013; 304: pp. R121–9.

[119] Buschur E, Kim C. Guidelines and interventions for obesity during pregnancy. *Int J Gynaecol Obstet*. 2012; 119: pp. 6–10.

[120] Poston L. Maternal obesity, gestational weight gain and diet as determinants of offspring long term health. *Best Pract Res Clin Endocrinol Metab*. 2012; 26: pp. 627–39.

[121] Khan IY, Taylor PD, Dekou V, Seed PT, Lakasing L, Graham D, Dominiczak AF, Hanson MA, Poston L. Gender-linked hypertension in offspring of lard-fed pregnant rats. *Hypertension*. 2003; 41: pp. 168–75.

[122] Samuelsson AM, Morris A, Igosheva N, Kirk SL, Pombo JM, Coen CW, Poston L, Taylor PD. Evidence for sympathetic origins of hypertension in juvenile offspring of obese rats. *Hypertension*. 2010; 55: pp. 76–82.

[123] Elahi MM, Cagampang FR, Mukhtar D, Anthony FW, Ohri SK, Hanson MA. Long-term maternal high-fat feeding from weaning through pregnancy and lactation predisposes offspring to hypertension, raised plasma lipids and fatty liver in mice. *Br J Nutr*. 2009; 102: pp. 514–9.

[124] Tsadok MA, Friedlander Y, Paltiel O, Manor O, Meiner V, Hochner H, Sagy Y, Sharon N, Yazdgerdi S, Siscovick D, Elchalal U. Obesity and blood pressure in 17-year-old offspring of mothers with gestational diabetes: insights from the Jerusalem Perinatal Study. *Exp Diabetes Res*. 2011; 2011: p. 906154.

[125] Aceti A, Santhakumaran S, Logan KM, Philipps LH, Prior E, Gale C, Hyde MJ, Modi N. The diabetic pregnancy and offspring blood pressure in childhood: a systematic review and meta-analysis. *Diabetologia*. 2012; 55: pp. 3114–27.

[126] Blondeau B, Joly B, Perret C, Prince S, Bruneval P, Lelièvre-Pégorier M, Fassot C, Duong Van Huyen JP. Exposure in utero to maternal diabetes leads to glucose intolerance and high blood pressure with no major effects on lipid metabolism. *Diabetes Metab*. 2011; 37: pp. 245–51.

[127] Katkhuda R, Peterson ES, Roghair RD, Norris AW, Scholz TD, Segar JL. Sex-specific programming of hypertension in offspring of late-gestation diabetic rats. *Pediatr Res*. 2012; 72: pp. 352–61.

[128] Fewtrell MS, Morley R, Abbott RA, Singhal A, Stephenson T, MacFadyen UM, Clements H, Lucas A. Catch-up growth in small-for-gestational-age term infants: a randomized trial. *Am J Clin Nutr*. 2001; 74: pp. 516–23.

[129] Leunissen RW, Kerkhof GF, Stijnen T, Hokken-Koelega AC. Effect of birth size and catch-up growth on adult blood pressure and carotid intima-media thickness. *Horm Res Paediatr.* 2012; 77: pp. 394–401.

[130] Boubred F, Daniel L, Buffat C, Feuerstein JM, Tsimaratos M, Oliver C, Dignat-George F, Lelièvre-Pégorier M, Simeoni U. Early postnatal overfeeding induces early chronic renal dysfunction in adult male rats. *Am J Physiol Renal Physiol.* 2009; 297: pp. F943–51.

[131] Ikenasio-Thorpe BA, Breier BH, Vickers MH, Fraser M. Prenatal influences on susceptibility to diet-induced obesity are mediated by altered neuroendocrine gene expression. *J Endocrinol.* 2007; 193: pp. 31–7.

[132] Woods LL, Ingelfinger JR, Rasch R. Modest maternal protein restriction fails to program adult hypertension in female rats. *Am J Physiol Regul Integr Comp Physiol.* 2005; 289: pp. R1131–6.

[133] Moritz KM, Mazzuca MQ, Siebel AL, Mibus A, Arena D, Tare M, Owens JA, Wlodek ME. Uteroplacental insufficiency causes a nephron deficit, modest renal insufficiency but no hypertension with ageing in female rats. *J Physiol.* 2009; 587: pp. 2635–46.

[134] Tang L, Carey LC, Bi J, Valego N, Sun X, Deibel P, Perrott J, Figueroa JP, Chappell MC, Rose JC. Gender differences in the effects of antenatal betamethasone exposure on renal function in adult sheep. *Am J Physiol Regul Integr Comp Physiol.* 2009; 296: pp. R309–17.

[135] Woods LL, Weeks DA, Rasch R. Programming of adult blood pressure by maternal protein restriction: role of nephrogenesis. *Kidney Int.* 2004; 65: pp. 1339–48.

[136] Musha Y, Itoh S, Hanson MA, Kinoshita K. Does estrogen affect the development of abnormal vascular function in offspring of rats fed a low-protein diet in pregnancy? *Pediatr Res.* 2006; 59: pp. 784–9.

[137] Intapad S, Tull FL, Brown AD, Dasinger JH, Ojeda NB, Fahling JM, Alexander BT. Renal denervation abolishes the age-dependent increase in blood pressure in female intrauterine growth-restricted rats at 12 months of age. *Hypertension.* 2013; 61: pp. 828–34.

[138] Gilbert JS, Nijland MJ. Sex differences in the developmental origins of hypertension and cardiorenal disease. *Am J Physiol Regul Integr Comp Physiol.* 2008; 295: pp. R1941–52.

[139] Huffman DM, Barzilai N. Contribution of adipose tissue to health span and longevity. *Interdiscip Top Gerontol.* 2010; 37: pp. 1–19.

[140] Chen W, Srinivasan SR, Berenson GS. Amplification of the association between birth weight and blood pressure with age: the Bogalusa Heart Study. *J Hypertens.* 2010; 28: pp. 2046–52.

[141] Virdis A, Bruno RM, Neves MF, Bernini G, Taddei S, Ghiadoni L. Hypertension in the elderly: an evidence-based review. *Curr Pharm Des.* 2011; 17: pp. 3020–31.

[142] Ojeda NB, Grigore D, Robertson EB, Alexander BT. Estrogen protects against increased

blood pressure in postpubertal female growth restricted offspring. *Hypertension.* 2007; 50: pp. 679–85.

[143] Salazar F, Reverte V, Saez F, Loria A, Llinas MT, Salazar FJ. Age- and sodium-sensitive hypertension and sex-dependent renal changes in rats with a reduced nephron number. *Hypertension.* 2008; 51: pp. 1184–9.

[144] Reverte V, Tapia A, Baile G, Gambini J, Gíménez I, Llinas MT, Salazar FJ. Role of angiotensin II in arterial pressure and renal hemodynamics in rats with altered renal development: age- and sex-dependent differences. *Am J Physiol Renal Physiol.* 2013; 304: pp. F33–40.

[145] Bertram JF, Douglas-Denton RN, Diouf B, Hughson MD, Hoy WE. Human nephron number: implications for health and disease. *Pediatr Nephrol.* 2011; 26: pp. 1529–33.

[146] Luyckx VA, Brenner BM. Low birth weight, nephron number, and kidney disease. *Kidney Int Suppl.* 2005; 97: pp. S68–77.

[147] Gossmann J, Wilhelm A, Kachel HG, Jordan J, Sann U, Geiger H, Kramer W, Scheuermann EH. Long-term consequences of live kidney donation follow-up in 93% of living kidney donors in a single transplant center. *Am J Transplant.* 2005; 5: pp. 2417–24.

[148] Woods LL. Neonatal uninephrectomy causes hypertension in adult rats. *Am J Physiol.* 1999 276: pp. R974–8.

[149] Moritz KM, Wintour EM, Dodic M. Fetal uninephrectomy leads to postnatal hypertension and compromised renal function. *Hypertension.* 2002; 39: pp. 1071–6.

[150] Rothermund L, Nierhaus M, Fialkowski O, Freese F, Ibscher R, Mieschel S, Kossmehl P, Grimm D, Wehland M, Kreutz R. Genetic low nephron number hypertension is associated with dysregulation of the hepatic and renal insulin-like growth factor system during nephrogenesis. *J Hypertens.* 2006; 24: pp. 1857–64.

[151] Stelloh C, Allen KP, Mattson DL, Lerch-Gaggl A, Reddy S, El-Meanawy A. Prematurity in mice leads to reduction in nephron number, hypertension, and proteinuria. *Transl Res.* 2012; 159: pp. 80–9.

[152] Hall JE. *Guyton and Hall Textbook of Physiology.* Twelfth Edition. Saunders, Elsevier, Philadelphia, PA United States. 2011.

[153] Vehaskari VM. Prenatal programming of kidney disease. *Curr Opin Pediatr.* 2010; 22: pp. 176–82.

[154] Ojeda NB, Grigore D, Yanes LL, Iliescu R, Robertson EB, Zhang H, Alexander BT. Testosterone contributes to marked elevations in mean arterial pressure in adult male intrauterine growth restricted offspring. *Am J Physiol Regul Integr Comp Physiol.* 2007; 292: pp. R758–63.

[155] Woods LL, Morgan TK, Resko JA. Castration fails to prevent prenatally programmed hypertension in male rats. *Am J Physiol Regul Integr Comp Physiol.* 2010; 298: pp. R1111–16.

[156] Wang H, Gallinat S, Li HW, Sumners C, Raizada MK, Katovich MJ. Elevated blood pres-

sure in normotensive rats produced by 'knockdown' of the angiotensin type 2 receptor. *Exp Physiol.* 2004; 89: pp. 313–22.

[157] Manning J, Vehaskari VM. Low birth weight-associated adult hypertension in the rat. *Pediatr Nephrol.* 2001; 16: pp. 417–22.

[158] Ceravolo GS, Franco MC, Carneiro-Ramos MS, Barreto-Chaves ML, Tostes RC, Nigro D, Fortes ZB, Carvalho MH. Enalapril and losartan restored blood pressure and vascular reactivity in intrauterine undernourished rats. *Life Sci.* 2007; 80: pp. 782–7.

[159] Wichi RB, Souza SB, Casarini DE, Morris M, Barreto-Chaves ML, Irigoyen MC. Increased blood pressure in the offspring of diabetic mothers. *Am J Physiol Regul Integr Comp Physiol.* 2005; 288: pp. R1129–33.

[160] Vehaskari VM, Stewart T, Lafont D, Soyez C, Seth D, Manning J. Kidney angiotensin and angiotensin receptor expression in prenatally programmed hypertension. *Am J Physiol Renal Physiol.* 2004; 287: pp. F262–7.

[161] Shaltout HA, Rose JC, Chappell MC, Diz DI. Angiotensin-(1-7) deficiency and baroreflex impairment precede the antenatal Betamethasone exposure-induced elevation in blood pressure. *Hypertension.* 2012; 59: pp. 453–8.

[162] Grigore D, Ojeda NB, Robertson EB, Dawson AS, Huffman C, Bourassa E, Speth RC, Brosnihan, KB, Alexander BT. Placental insufficiency results in temporal alterations in the renin angiotensin system in male hypertensive growth restricted offspring. *Am J Physiol Regul Integr Comp Physiol.* 2007; 293: pp. R804–11.

[163] Crowley SD, Coffman TM. Recent advances involving the renin–angiotensin system. *Exp Cell Res.* 2012; 318: pp. 1049–56.

[164] Santos RA, Ferreira AJ, Verano-Braga T, Bader M. Angiotensin-converting enzyme 2, angiotensin-(1-7) and Mas: new players of the renin-angiotensin system. *J Endocrinol.* 2013; 216: pp. R1–17.

[165] Ferrario CM. ACE2: more of Ang-(1-7) or less Ang II? *Curr Opin Nephrol Hypertens.* 2011; 20: pp. 1–6.

[166] Sampson AK, Moritz KM, Jones ES, Flower RL, Widdop RE, Denton KM. Enhanced angiotensin II type 2 receptor mechanisms mediate decreases in arterial pressure attributable o chronic low-dose angiotensin II in female rats. *Hypertension.* 2008; 52: pp. 666–71.

[167] Brosnihan KB, Hodgin JB, Smithies O, Maeda N, Gallagher P. Tissue-specific regulation of ACE/ACE2 and AT1/AT2 receptor gene expression by oestrogen in apolipoprotein E/oestrogen receptor-alpha knock-out mice. *Exp Physiol.* 2008; 93: pp. 658–64.

[168] Liu J, Ji H, Zheng W, Wu X, Zhu JJ, Arnold AP, Sandberg K. Sex differences in renal angiotensin converting enzyme 2 (ACE2) activity are 17β-oestradiol-dependent and sex chromosome-independent. *Biol Sex Differ.* 2010; 1(1):6. doi: 10.1186/2042-6410-1-6.

[169]   Ding Y, Sigmund CD. Androgen-dependent regulation of human angiotensinogen expression in KAP-hAGT transgenic mice. *Am J Physiol Renal Physiol.* 2001; 280: pp. F54–60.

[170]   Hilliard LM, Sampson AK, Brown RD, Denton KM. The "his and hers" of the renin-angiotensin system. *Curr Hypertens Rep.* 2013; 15: pp. 71–9.

[171]   McMullen S, Langley-Evans SC. Sex-specific effects of prenatal low-protein and carbenoxolone exposure on renal angiotensin receptor expression in rats. *Hypertension.* 2005; 46: pp. 1374–80.

[172]   Arnold AC, Gallagher PE, Diz DI. Brain renin-angiotensin system in the nexus of hypertension and aging. *Hypertens Res.* 2013; 36: pp. 5–13.

[173]   Dodic M, Abouantoun T, O'Connor A, Wintour EM, Moritz KM. Programming effects of short prenatal exposure to dexamethasone in sheep. *Hypertension.* 2002; 40: pp. 729–34.

[174]   Goyal R, Goyal D, Leitzke A, Gheorghe CP, Longo LD. Brain renin-angiotensin system: fetal epigenetic programming by maternal protein restriction during pregnancy. *Reprod Sci.* 2010; 17: pp. 227–38.

[175]   Pladys P, Lahaie I, Cambonie G, Thibault G, Lê NL, Abran D, Nuyt AM. Role of the brain and peripheral angiotensin II in hypertension and altered arterial baroreflex programmed during fetal life in the rat. *Pediatr Res.* 2004; 55: pp. 1042–9.

[176]   Ojeda NB, Royals TP, Black JT, Dasinger JH, Johnson JM, Alexander BT. Enhanced sensitivity to acute angiotensin II is testosterone dependent in adult male growth-restricted offspring. *Am J Physiol Regul Integr Comp Physiol.* 2010; 298: pp. R1421–7.

[177]   Ojeda NB, Intapad S, Royals TP, Black JT, Dasinger JH, Tull FL, Alexander BT. Hypersensitivity to acute ANG II in female growth-restricted offspring is exacerbated by ovariectomy. *Am J Physiol Regul Integr Comp Physiol.* 2011; 301: pp. R1199–205.

[178]   DiBona GF. Sympathetic nervous system and the kidney in hypertension. *Curr Opin Nephrol Hypertens.* 2002; 11: pp. 197–200.

[179]   IJzerman RG, Stehouwer CD, de Geus EJ, van Weissenbruch MM, Delemarre-van de Waal HA, Boomsma DI. Low birth weight is associated with increased sympathetic activity: dependence on genetic factors. *Circulation.* 2003; 108: pp. 566–71.

[180]   Weitz G, Deckert P, Heindl S, Struck J, Perras B, Dodt C. Evidence for lower sympathetic nerve activity in young adults with low birth weight. *J Hypertens.* 2003; 21: pp. 943–50.

[181]   Alexander BT, Hendon AE, Ferril G, Dwyer TM. Renal denervation abolishes hypertension in low birth weight offspring from pregnant rats with reduced uterine perfusion. *Hypertension.* 2005; 45: pp. 754–8.

[182]   Dagan A, Kwon HM, Dwarakanath V, Baum M. Effect of renal denervation on prenatal programming of hypertension and renal tubular transporter abundance. *Am J Physiol Renal Physiol.* 2008; 295: pp. F29–34.

[183] Wang X, Armando I, Upadhyay K, Pascua A, Jose PA. The regulation of proximal tubular salt transport in hypertension: an update. *Curr Opin Nephrol Hypertens.* 2009; 18: pp. 412–20.

[184] Dagan A, Gattineni J, Cook V, Baum M. Prenatal programming of rat proximal tubule Na+/H+ exchanger by dexamethasone. *Am J Physiol Regul Integr Comp Physiol.* 2007; 292: pp. R1230–R1235.

[185] Manning J, Beutler K, Knepper MA, Vehaskari VM. Upregulation of renal BSC1 and TSC in prenatally programmed hypertension. *Am J Physiol Renal Physiol.* 2002; 283: pp. F202–6.

[186] Dalziel SR, Parag V, Rodgers A, Harding JE. Cardiovascular risk factors at age 30 following pre-term birth. *Int J Epidemiol.* 2007; 36: pp. 907–15.

[187] Ligi I, Grandvuillemin I, Andres V, Dignat-George F, Simeoni U. Low birth weight infants and the developmental programming of hypertension: a focus on vascular factors. *Semin Perinatol.* 2010; 34: pp. 188–92.

[188] Zanardo V, Fanelli T, Weiner G, Fanos V, Zaninotto M, Visentin S, Cavallin F, Trevisanuto D, Cosmi E. Intrauterine growth restriction is associated with persistent aortic wall thickening and glomerular proteinuria during infancy. *Kidney Int.* 2011; 80: pp. 119–23.

[189] Gunes T, Akin MA, Canoz O, Coban D, Ozcan B, Kose M, Ozturk MA, Kurtoglu S. Aortic intima-media thickness in nicotine-exposed rat pups during gestation and lactation period. *Eur J Pediatr.* 2011; 170: pp. 1257–62.

[190] Hellström A, Dahlgren J, Marsál K, Ley D. Abnormal retinal vascular morphology in young adults following intrauterine growth restriction. *Pediatrics.* 2004; 113: pp. e77–80.

[191] Payne JA, Alexander BT, Khalil RA. Reduced endothelial vascular relaxation in growth-restricted offspring of pregnant rats with reduced uterine perfusion. *Hypertension.* 2003; 42: pp. 768–74.

[192] Segar EM, Norris AW, Yao JR, Hu S, Koppenhafer SL, Roghair RD, Segar JL, Scholz TD. Programming of growth, insulin resistance and vascular dysfunction in offspring of late gestation diabetic rats. *Clin Sci (Lond).* 2009; 117: pp. 129–38.

[193] Morton JS, Rueda-Clausen CF, Davidge ST. Flow-mediated vasodilation is impaired in adult rat offspring exposed to prenatal hypoxia. *J Appl Physiol.* 2011; 110: pp. 1073–82.

[194] Anderson CM, Lopez F, Zimmer A, Benoit JN. Placental insufficiency leads to developmental hypertension and mesenteric artery dysfunction in two generations of Sprague–Dawley rat offspring. *Biol Reprod.* 2006; 74: pp. 538–44.

[195] Franco Mdo C, Ponzio BF, Gomes GN, Gil FZ, Tostes R, Carvalho MH, Fortes ZB. Micronutrient prenatal supplementation prevents the development of hypertension and vascular endothelial damage induced by intrauterine malnutrition. *Life Sci.* 2009; 85: pp. 327–33.

[196] Montezano AC, Touyz RM. Oxidative stress, Noxs, and hypertension: experimental evidence and clinical controversies. *Ann Med.* 2012; 44: pp. S2–16.

[197]   Negi R, Pande D, Kumar A, Khanna RS, Khanna HD. Evaluation of biomarkers of oxidative stress and antioxidant capacity in the cord blood of preterm low birth weight neonates. *J Matern Fetal Neonatal Med.* 2012; 25: pp. 1338–41.

[198]   Cambonie G, Comte B, Yzydorczyk C, Ntimbane T, Germain N, Lê NL, Pladys P, Gauthier C, Lahaie I, Abran D, Lavoie JC, Nuyt AM. Antenatal antioxidant prevents adult hypertension, vascular dysfunction, and microvascular rarefaction associated with in utero exposure to a low-protein diet. *Am J Physiol Regul Integr Comp Physiol.* 2007; 292: pp. R1236–45.

[199]   Franco MC, Kawamoto EM, Gorjão R, Rastelli VM, Curi R, Scavone C, Sawaya AL, Fortes ZB, Sesso R. Biomarkers of oxidative stress and antioxidant status in children born small for gestational age: evidence of lipid peroxidation. *Pediatr Res.* 2007; 62: pp. 204–8.

[200]   Ojeda NB, Hennington BS, Williamson DT, Hill ML, Betson NE, Sartori-Valinotti JC, Reckelhoff JF, Royals TP, Alexander BT. Oxidative stress contributes to sex differences in blood pressure in adult growth-restricted offspring. *Hypertension.* 2012:60: pp. 114–22.

[201]   Stewart T, Jung FF, Manning J, Vehaskari VM. Kidney immune cell infiltration and oxidative stress contribute to prenatally programmed hypertension. *Kidney Int.* 2005; 68: pp. 2180–8.

[202]   Franco Mdo C, Akamine EH, Aparecida de Oliveira M, Fortes ZB, Tostes RC, Carvalho MH, Nigro D. Vitamins C and E improve endothelial dysfunction in intrauterine-undernourished rats by decreasing vascular superoxide anion concentration. *J Cardiovasc Pharmacol.* 2003; 42: pp. 211–7.

[203]   Lopez-Ruiz A, Sartori-Valinotti J, Yanes LL, Iliescu R, Reckelhoff JF. Sex differences in control of blood pressure: role of oxidative stress in hypertension in females. *Am J Physiol Heart Circ Physiol.* 2008; 295: pp. H466–74.

[204]   Dantas AP, Franco Mdo C, Silva-Antonialli MM, Tostes RC, Fortes ZB, Nigro D, Carvalho MH. Gender differences in superoxide generation in microvessels of hypertensive rats: role of NAD(P)H-oxidase. *Cardiovasc Res.* 2004; 61: pp. 22–9.

[205]   Bhatia K, Elmarakby AA, El-Remessy AB, Sullivan JC. Oxidative stress contributes to sex differences in angiotensin II-mediated hypertension in spontaneously hypertensive rats. *Am J Physiol Regul Integr Comp Physiol.* 2012; 302: pp. R274–82.

[206]   Welch WJ, Wilcox CS. AT_1 receptor antagonist combats oxidative stress and restores nitric oxide signaling in SHR. *Kidney Int.* 2001; 59: pp. 1257–63.

[207]   Franco Mdo C, Akamine EH, Di Marco GS, Casarini DE, Fortes ZB, Tostes RC, Carvalho MH, Nigro D. NADPH oxidase and enhanced superoxide generation in intrauterine undernourished rats: involvement of the renin-angiotensin system. *Cardiovasc Res.* 2003 59: pp. 767–75.

[208] Seckl JR. Prenatal glucocorticoids and long-term programming. *Eur J Endocrinol.* 2004; 151: pp. U49–62.

[209] Woods LL, Weeks DA. Prenatal programming of adult blood pressure: role of maternal corticosteroids. *Am J Physiol Regul Integr Comp Physiol.* 2005; 289: pp. R955–62.

[210] Dy J, Guan H, Sampath-Kumar R, Richardson BS, Yang K. Placental 11beta-hydroxysteroid dehydrogenase type 2 is reduced in pregnancies complicated with idiopathic intrauterine growth restriction: evidence that this is associated with an attenuated ratio of cortisone to cortisol in the umbilical artery. *Placenta.* 2008; 29: pp. 193–200.

[211] Langley-Evans SC, Welham SJ, Sherman RC, Jackson AA. Weanling rats exposed to maternal low-protein diets during discrete periods of gestation exhibit differing severity of hypertension. *Clin Sci (Lond).* 1996; 91: pp. 607–15.

[212] Chen M, Wang T, Liao ZX, Pan XL, Feng YH, Wang H. Nicotine-induced prenatal over-exposure to maternal glucocorticoid and intrauterine growth retardation in rat. *Exp Toxicol Pathol.* 2007; 59: pp. 245–51.

[213] Langley-Evans SC. Hypertension induced by foetal exposure to a maternal low-protein diet, in the rat, is prevented by pharmacological blockade of maternal glucocorticoid synthesis. *J Hypertens.* 1997; 15: pp. 537–44.

[214] Baserga M, Kaur R, Hale MA, Bares A, Yu X, Callaway CW, McKnight RA, Lane RH. Fetal growth restriction alters transcription factor binding and epigenetic mechanisms of renal 11beta-hydroxysteroid dehydrogenase type 2 in a sex-specific manner. *Am J Physiol Regul Integr Comp Physiol.* 2010; 299: pp. R334–42.

[215] Webster AL, Yan MS, Marsden PA. Epigenetics and cardiovascular disease. *Can J Cardiol.* 2013; 29: pp. 46–57.

[216] Bogdarina I, Welham S, King PJ, Burns SP, Clark AJ. Epigenetic modification of the renin-angiotensin system in the fetal programming of hypertension. *Circ Res.* 2007; 100: pp. 520–6.

[217] Pham TD, MacLennan NK, Chiu CT, Laksana GS, Hsu JL, Lane RH. Uteroplacental insufficiency increases apoptosis and alters p53 gene methylation in the full-term IUGR rat kidney. *Am J Physiol Regul Integr Comp Physiol.* 2003; 285: pp. R962–70.

[218] Lillycrop KA, Slater-Jefferies JL, Hanson MA, Godfrey KM, Jackson AA, Burdge GC. Induction of altered epigenetic regulation of the hepatic glucocorticoid receptor in the offspring of rats fed a protein-restricted diet during pregnancy suggests that reduced DNA methyltransferase-1 expression is involved in impaired DNA methylation and changes in histone modifications. *Br J Nutr.* 2007; 97: pp. 1064–73.

[219] Burdge GC, Slater-Jefferies J, Torrens C, Phillips ES, Hanson MA, Lillycrop KA. Dietary protein restriction of pregnant rats in the F0 generation induces altered methylation of

hepatic gene promoters in the adult male offspring in the F1 and F2 generations. *Br J Nutr.* 2007; 97: pp. 435–9.

[220]   Ponzio BF, Carvalho MH, Fortes ZB, do Carmo Franco M. Implications of maternal nutrient restriction in transgenerational programming of hypertension and endothelial dysfunction across F1–F3 offspring. *Life Sci.* 2012; 90: pp. 571–7.

[221]   Harrison M, Langley-Evans SC. Intergenerational programming of impaired nephrogenesis and hypertension in rats following maternal protein restriction during pregnancy. *Br J Nutr.* 2009; 101: pp. 1020–30.

[222]   Torrens C, Poston L, Hanson MA. Transmission of raised blood pressure and endothelial dysfunction to the F2 generation induced by maternal protein restriction in the F0, in the absence of dietary challenge in the F1 generation. *Br J Nutr.* 2008; 100: pp. 760–6.

[223]   Jennings BJ, Ozanne SE, Dorling MW, Hales CN. Early growth determines longevity in male rats and may be related to telomere shortening in the kidney. *FEBS Lett.* 1999; 448, pp. 4–8.

[224]   Gomez DE, Armando RG, Farina HG, Menna PL, Cerrudo CS, Ghiringhelli PD, Alonso DF. Telomere structure and telomerase in health and disease (review). *Int J Oncol.* 2012; 41: pp. 1561–9.

[225]   Tarry-Adkins JL, Martin-Gronert MS, Fernandez-Twinn DS, Hargreaves I, Alfaradhi MZ, Land JM, Aiken CE, Ozanne SE. Poor maternal nutrition followed by accelerated postnatal growth leads to alterations in DNA damage and repair, oxidative and nitrosative stress, and oxidative defense capacity in rat heart. *FASEB J.* 2013; 27: pp. 379–90.

[226]   Tarry-Adkins JL, Chen JH, Smith NS, Jones RH, Cherif H, Ozanne SE. Poor maternal nutrition followed by accelerated postnatal growth leads to telomere shortening and increased markers of cell senescence in rat islets. *FASEB J.* 2009; 23: pp. 1521–8.

[227]   Kajantie E, Pietiläinen KH, Wehkalampi K, Kananen L, Räikkönen K, Rissanen A, Hovi P, Kaprio J, Andersson S, Eriksson JG, Hovatta I. No association between body size at birth and leucocyte telomere length in adult life—evidence from three cohort studies. *Int J Epidemiol.* 2012; 1: pp. 1400–8.

[228]   Entringer S, Epel ES, Kumsta R, Lin J, Hellhammer DH, Blackburn EH, Wüst S, Wadhwa PD. Stress exposure in intrauterine life is associated with shorter telomere length in young adulthood. *Proc Natl Acad Sci U S A.* 2011; 108: pp. E513–8.

[229]   Entringer S, Epel ES, Lin J, Buss C, Shahbaba B, Blackburn EH, Simhan HN, Wadhwa PD. Maternal psychosocial stress during pregnancy is associated with newborn leukocyte telomere length. *Am J Obstet Gynecol.* 2013; 208: pp. 134.e1–7.

[230]   Metcalfe NB, Monaghan P. Compensation for a bad start: grow now, pay later? *Trends Ecol Evol.* 2001; 16: pp. 254–60.

[231] Victora CG, Barros FC, Horta BL, Martorell R. Short-term benefits of catch-up growth for small for gestational age infants. *Int J Epidemiol.* 2001; 30: pp. 1325–30.

[232] Lau C, Rogers JM, Desai M, Ross MG. Fetal programming of adult disease: implications for prenatal care. *Obstet Gynecol.* 2011; 117: pp. 978–85.

[233] Ben-Shlomo Y, McCarthy A, Hughes R, Tilling K, Davies D, Smith GD. Immediate postnatal growth is associated with blood pressure in young adulthood: the Barry Caerphilly Growth Study. *Hypertension.* 2008; 52: pp. 638–44.

[234] Coupé B, Grit I, Darmaun D, Parnet P. The timing of catch-up growth affects metabolism and appetite regulation in male rats born with intrauterine growth restriction. *Am J Physiol Regul Integr Comp Physiol.* 2009; 297: pp. R813–24.

[235] Barrett-Connor E. Menopause, atherosclerosis, and coronary artery disease. *Curr Opin Pharmacol.* 2013. doi: pii: S1471-4892(13)00009-X. 10.1016/j.coph.2013.01.005.

[236] Vaidya D, Becker DM, Bittner V, Mathias RA, Ouyang P. Ageing, menopause, and ischaemic heart disease mortality in England, Wales, and the United States: modelling study of national mortality data. *Br Med J.* 2011; 343: p. d5170.

[237] Yarbrough DE, Barrett-Connor E, Kritz-Silverstein D, Wingard DL. Birth weight, adult weight, and girth as predictors of the metabolic syndrome in postmenopausal women: the Rancho Bernardo Study. *Diabetes Care.* 1998; 21: pp. 1652–8.

[238] Wellons M, Ouynag P, Schreiner PJ, Herrington DM, Vaidya D. Early menopause predicts future coronary heart disease and stroke: the multi-ethnic study of atherosclerosis. *Menopause.* 2012; 19: pp. 1081–7.

[239] Luborsky JL, Meyer P, Sowers MF, Gold EB, Santoro N. Premature menopause in a multi-ethnic population study of the menopause transition. *Hum Reprod.* 2003; 18: pp. 199–206.

[240] North American Menopause Society. *Menopause Practice: A Clinician's Guide.* 3rd ed. North American Menopause Society, Cleveland, OH, 2007.

[241] Torgerson DJ, Thomas RE, Reid DM. Mothers and daughters menopausal ages: is there a link? *Eur J Obstet Gynecol Reprod Biol.* 1997; 74: pp. 63–6.

[242] Bromberger JT, Matthews KA, Kuller LH, Wing RR, Meilahn EN, Plantinga P. Prospective study of the determinants of age at menopause. *Am J Epidemiol.* 1997; 145: pp. 124–33.

[243] Steiner AZ, D'Aloisio AA, DeRoo LA, Sandler DP, Baird DD. Association of intrauterine and early-life exposures with age at menopause in the Sister Study. *Am J Epidemiol.* 2010; 172: pp. 140–8.

[244] Cresswell JL, Egger P, Fall CH, Osmond C, Fraser RB, Barker DJ. Is the age of menopause determined in-utero? *Early Hum Dev.* 1997; 49: pp. 143–8.

[245] McKnight JR, Satterfield MC, Li X, Gao H, Wang J, Li D, Wu G. Obesity in pregnancy: problems and potential solutions. *Front Biosci (Elite Ed).* 2011; 3: pp. 442–52.

[246]   Dempsey JC, Williams MA, Luthy DA, Emanuel I, Shy K. Weight at birth and subsequent risk of preeclampsia as an adult. *Am J Obstet Gynecol.* 2003; 189: pp. 494–500.

[247]   Gallo LA, Tran M, Moritz KM, Jefferies AJ, Wlodek ME. Pregnancy in aged rats that were born small: cardiorenal and metabolic adaptations and second-generation fetal growth. *FASEB J.* 2012; 26: pp. 4337–47.

[248]   Gallo LA, Denton KM, Moritz KM, Tare M, Parkington HC, Davies M, Tran M, Jefferies AJ, Wlodek ME. Long-term alteration in maternal blood pressure and renal function after pregnancy in normal and growth-restricted rats. *Hypertension.* 2012; 60: pp. 206–213.

[249]   Tran M, Gallo LA, Wadley GD, Jefferies AJ, Moritz KM, Wlodek ME. Effect of pregnancy for females born small on later life metabolic disease risk. *PLoS One.* 2012; 7: p. e45188.

[250]   Chauhan SP, Ananth CV. Induction of labor in the United States: a critical appraisal of appropriateness and reducibility. *Semin Perinatol.* 2012; 36: pp. 336–43.

[251]   Kozhimannil KB, Law MR, Virnig BA. Cesarean delivery rates vary tenfold among US hospitals; reducing variation may address quality and cost issues. *Health Aff (Millwood).* 2013; 32: pp. 527–35.

[252]   Mensah GA, Mokdad AH, Ford ES, Greenlund KJ, Croft JB. State of disparities in cardiovascular health in the United States. *Circulation.* 2005; 111: pp. 1233–41.

[253]   Romero CX, Romero TE, Shlay JC, Ogden LG, Dabelea D. Changing trends in the prevalence and disparities of obesity and other cardiovascular disease risk factors in three racial/ethnic groups of USA adults. *Adv Prev Med.* 2012; 2012: p. 172423.

[254]   US Renal Data System, USRDS 2008 Annual Data Report, The National Institutes of Health, National Institute of Diabetes and Digestive and Kidney Diseases, Bethesda, MD, 2008.

[255]   Feinstein M, Ning H, Kang J, Bertoni A, Carnethon M, Lloyd-Jones DM. Racial differences in risks for first cardiovascular events and noncardiovascular death: the Atherosclerosis Risk in Communities study, the Cardiovascular Health Study, and the Multi-Ethnic Study of Atherosclerosis. *Circulation.* 2012; 126: pp. 50–9.

[256]   Fox CS, Muntner P. Trends in diabetes, high cholesterol, and hypertension in chronic kidney disease among U.S. adults: 1988–1994 to 1999–2004. *Diabetes Care.* 2008: pp. 1337–42.

[257]   Hamilton BE, Hoyert DL, Martin JA, Strobino DM, Guyer B. Annual summary of vital statistics: 2010–2011. *Pediatrics.* 2013; 131: pp. 548–58.

[258]   Ruggenenti P, Cravedi P, Remuzzi G. Mechanisms and treatment of CKD. *J Am Soc Nephrol.* 2012; 23: pp. 1917–28.

[259]   Lim K, Lombardo P, Schneider-Kolsky M, Hilliard L, Denton KM, Black MJ. Induction of hyperglycemia in adult intrauterine growth-restricted rats: effects on renal function. *Am J Physiol Renal Physiol.* 2011; 301: pp. F288–94.

[260] Ojeda NB. Low birth weight increases susceptibility to renal injury in a rat model of mild ischemia-reperfusion. *Am J Physiol Renal Physiol.* 2011; 301: pp. F420–6.

[261] Bogdarina I, Haase A, Langley-Evans S, Clark AJ. Glucocorticoid effects on the programming of AT1b angiotensin receptor gene methylation and expression in the rat. *PLoS One.* 2010; 5: pp. e9237.

[262] Hall JE. Control of sodium excretion by .angiotensin I86;250:I: intrarenal mechanisms and blood pressure regulation. *Am J Physiol.* 1986; 250: pp. R960–72.

[263] Gomez RA, Lynch KR, Chevalier RL, Wilfong N, Everett A, Carey RM, Peach MJ. Renin and angiotensinogen gene expression in maturing rat kidney. *Am J Physiol Renal Fluid Electrolyte Physiol.* 1988; 254: pp. F582–7.

[264] Davisson RL, Ding Y, Stec DE, Catterall JF, Sigmund CD. Novel mechanism of hypertension revealed by cell-specific targeting of human angiotensinogen in transgenic mice. *Physiol Genomics.* 1999; 1: pp. 3–9.

[265] Whaley-Connell A, Sowers JR. Oxidative stress in the cardiorenal metabolic syndrome. *Curr Hypertens Rep.* 2012; 14: pp. 360–5.

[266] Ibáñez L, Potau N, Enriquez G, de Zegher F. Reduced uterine and ovarian size in adolescent girls born small for gestational age. *Pediatr Res.* 2000; 47:575–7.

Author Biography

Barbara T. Alexander, Ph.D., is an Associate Professor of Physiology at the University of Mississippi Medical Center. She received her B.S. degree in zoology from Mississippi State University and her Ph.D. degree in biochemistry from the University of Mississippi. She conducted her postdoctoral training with Dr. Joey P. Granger at the University of Mississippi Medical Center where her research centered on the renal mechanisms linking placental ischemia with hypertension during pregnancy. Her current research focus involves investigation into the mechanisms by which adverse influences during early life program later chronic health with a specific focus on sex differences in the fetal response to programmed cardiovascular risk. Dr. Alexander has received funding from the American Heart Association and is currently funded by the National Institutes of Health. She is a Fellow for the Council for High Blood Pressure Research of the American Heart Association and is an active member of the American Physiological Society.